Autism:
A guide to understanding and helping your child.

Andrew Brereton

First published in 2007 by Snowdrop.

© Andrew Brereton. All rights reserved. No part of this publication may be reproduced, stored in a retrieval system, transmitted or utilised in any form or by any means, electronic, mechanical, photocopying, recording or otherwise, without permission in writing from the publisher..

ISBN. 978-0-9557309-1-7

Autism: A guide to understanding and helping your child.

Snowdrop

This book is for information only. Parents should not attempt any of the techniques contained within this book with their children until they have undergone a full developmental assessment by Snowdrop. Parents attempting any techniques contained within this book prior to a full developmental evaluation do so at their own risk. Snowdrop takes no responsibility for any child until the organisation has assessed the child and prescribed an appropriate programme of developmental activities.

For all the children who do not perceive the world as we do.

Contents

Introduction	9
Disclosure of diagnosis	15
What is Autism?	19
Reaction and Adjustment	23
Problems of sensory perception	27
Vision	31
Auditory development	37
Tactile development	45
Language development	51
Mobility development	61
Fine motor development	67
Social development and the sense of self	71
Treatment	77
Treating sensory distortions	81
Treating language ifficulties	95
Treating mobility problems	103
Learning difficulties	109
Sleep	117
Practical Issues	123
Epilepsy	125
Parent, professional and family relations	133
All you need is love	141
Where do we go from here	143
References	145

Introduction

If you are reading this book, your child has probably received a diagnosis of autism, or Asperger's syndrome, or pervasive developmental disorder, or whatever new label they have recently thought up and you are probably unaware of just what these labels mean.

I can already hear the howls of professional indignation at the fact that I have labelled autism as a brain-injury. Autism has a genetic component I hear them say and they are correct, ***in some children.*** However, I have seen too many autistic type symptoms in children who have suffered brain-injuries to ignore the fact that autism can have an environmental component too! – That environmental component is brain-injury, whether that brain injury is caused by oxygen starvation at birth, by factors during pregnancy, by head injury, by the brain for some reason adopting a faulty wiring pattern, or by any other cause!

In any case, if one were to consider the impact of genes on autism, one would have to conclude that the expression of the genes associated with autism, causes neurological dysfunction – in other words it is another route to brain-injury!

You will understandably be distressed by this; - You are also likely to be suffering from shock, confusion, bewilderment and even anger. These feelings will eventually pass, (apart from the bewilderment, which from personal experience is likely to remain). I speak from the experience of being a parent to a child who suffered profound brain-injuries and who displayed many autistic spectrum symptoms. Speaking personally, although I have gained qualifications in neuropsychology and child development and studied in various clinics both internationally and in the UK, which offer many varied forms of treatment, my own child remains my most astute tutor, (as your children will be for you). All I have learned and experienced during this time has led me to a greater understanding of the immense difficulties faced by children and their families.

Autism is a formidable enemy, as is any other pattern of brain-injury. If it is unchallenged, it prevents the development of the child's functional capability, eats away at the family's sovereignty and is a constant threat to the child's health and wellbeing. The failures of social development, communication and imagination, which children with autism experience, (alongside many other difficulties), do not need to be unchallenged as there is a wealth of knowledge available, which can be utilised to assist the child's development.

It is a fact that the brain has a high degree of plasticity. – What does this mean? Well it means that brain cells are able to change their function or organisation due to environmental experience. (Casasanto, 1998). This means, if we are able to provide the child with an environment, which has the optimum level of appropriate stimulation within it, we should be able to encourage his development. Indeed, the work many researchers, demonstrates very clearly to us that the brain retains a high degree of plasticity, not only throughout our young life, but well into old age!

This moves us to a principle of treatment; - The provision of a developmental environment designed to stimulate and to maximise neural plasticity!

This book has a sharp focus. It is intended as a guide to parental understanding; - and with this in mind I have attempted to keep technical language to a minimum. However, sometimes this is impossible and so where I have been forced to become technical, I have made efforts to provide explanations. I know from personal experience that what parents need is clear, concise information, (as free as possible of technical jargon), which strikes home as true as the truest arrow.

Another reason for writing the book is to coincide with my establishing a child development consultancy called 'Snowdrop.' The idea of Snowdrop is that families who are not satisfied with the developmental progress, which their child is currently achieving, can access treatment designed to maximise that progress. This comes in the form of a prescription of a programme of developmental activities, which can be carried out at home.

The book has attempted to achieve a situation, where it examines autistic type conditions and their major effects upon child development and presents the information in a clear, accessible and informative manner. I have attempted to pull together the best ideas from the worlds of psychology, neuropsychology and neurophysiology and present them alongside my own theories, in an attempt to improve children's developmental prospects and in turn help the families involved. Snowdrop's treatment ideas have been developed from the work of several eminent psychologists / neurologists of the past and a few of the present day.

There are many who have been influential on my thinking: - Luria, Kolb, Wishaw, Rogoff and Wood to name just a few. However, the man who has exercised the greatest level of influence upon my philosophy is the great Russian psychologist Lev Vygotsky. Vygotsky's work on the relationship between children's' learning and development is only just beginning to be embraced by a Western culture, which has seemed infatuated with behaviourism and in latter years, the theories of Jean Piaget. Indeed, despite the fact that behaviourist techniques are of only limited use with the most severely brain-injured children, it is sad to see just how many behaviourist techniques are still applied in schools in the twenty-first century. This is despite Chomsky's demolition of behaviourist philosophy many years earlier.

Behaviourism is the belief that to understand an individual, one only has to observe his behaviour and that a person's behaviour can be influenced through the utilisation of the laws of cause and effect.

Jean Piaget was a Swiss psychologist whose work in the mid twentieth century focussed on the proposal that children's intellectual development occurs in definite stages, He asserted that a process called 'maturation' plays a key role in children's increasing developmental capabilities: He believed therefore that children were incapable of moving into higher developmental levels until their brains had matured sufficiently, (Basically, until they had reached the correct age). Subsequent research has undermined many of Piaget's ideas. Unfortunately, even though they have been undermined, Piaget's ideas have been used as the basis for organising the school curriculum in many Western societies.

By contrast, the ideas of Lev Vygotsky concerning how children learn and develop were radical in their opposition to Piaget's theories. Vygotsky believed that children learn, indeed all of us learn most effectively in social situations. He claimed that learning takes place in what he termed the ***zone of proximal development.*** He also claimed that the most effective learning is supported by a more skilled partner, who gives the learner a type of cognitive support, which we now know as ***scaffolding.*** He proposed that adults give this support, known as scaffolding, to children whilst they are learning. There is a great deal of evidence to suggest that this is indeed the case.

We are fortunate that Vygotsky's theories can be successfully adapted and applied to the treatment of brain-injured children and indeed his work leads us to another principle of treatment. - Learning leads development!

Vygotsky's proposition of being able to increase children's developmental capability through instruction in this way, is in total opposition to the position of Piaget who believed that before specific ages, children are simply not capable of understanding certain concepts.

> *The zone of proximal development and scaffolding: - Bruner coined the term 'scaffolding' to define just the type of supported learning observed in children's homes; learning which takes place through what Vygotsky termed the 'zone of proximal development.' (Wood et al, 1976). This is a situation where the learning, which is being assisted through the guidance and support of a parent, is slightly in advance of the child's current capability to carry out the task successfully alone. As the child's ability to carry out the task increases, the support and guidance (the scaffold), given is gradually decreased, leading to a situation where the new ability becomes internalised as part of the child's developmental capability.*

The concept of scaffolding was later, further refined by Rogoff, who described such supported learning as a process of **guided participation.**

> *Guided participation: - a type of scaffolding which acknowledges that children are not passive receivers of instruction, but are active participants in their own learning.*

My philosophy regarding treatment has been developed over many years, studying and working with individual children, alongside academic research during periods of study at a wide range of clinics all over the world. I have also examined conventional approaches to the treatment and management of various types of brain-injury, as employed by medical / health establishments in different countries and incorporated beneficial aspects of these treatment regimes alongside my own.

Snowdrop aims to be an eclectic organisation and does see immense value in working alongside the established system of treatment whilst at the same time advancing our own ideas. Therefore, although Snowdrop's approach is radical and 'learning based,' it does not aim to be outside of or 'anti' the establishment'.

Alongside my selective praise for aspects of the medical approach to treatment, it has to be said that there are also some useful and valid alternative approaches to the treatment of brain-injury. I believe that all of these positive contributions to child development can be merged and incorporated into a more heuristic, wide-ranging basis for treatment

I also have experience of working in schools, working with children who have special educational needs, attempting where possible to ensure their needs are met and that their opportunities for learning and development in the classroom are maximised. However, despite what it seems Government policy in the UK and elsewhere would have us believe, many children are simply not suitable candidates for the environment of a classroom. Indeed, due to their extremely complex problems, they are not suitable candidates to be comfortable in almost any environment. Many children who suffer autism fall into this category.

What this book tries to do is to convey the message that there is hope for all children. No child is ineducable and no child should be denied the possibility of having his developmental prospects improved. All we have to do is to provide an environment, which is designed and structured to be at just the right level to facilitate learning and neurological restructuring.

Most of all we need to remember two vital messages, which should underpin any treatment philosophy.

- ***The brain is plastic.***
- ***Learning leads development.***

Disclosure of Diagnosis

Most children are not diagnosed with autism until they are 2 years old or over. A child might be diagnosed as having brain injuries before then but generally, even so, most doctors will not use the term 'autism' until much later. Whether the disagnosis is brain injuries at birth or shortly after, or whether it is autism at 2 years old or more is in one sense irrelevant. Most people cannot even imagine the horror of being informed that their child is different; - their child has problems. They cannot imagine the trauma of being informed that he will not pass the milestones, which other children so easily pass; he will not achieve what comes so easily to others. This is a harrowing experience. At some point in the past, whether recent or distant, you will have experienced the horror of this situation. I personally experienced it 23 years ago when a consultant paediatrician calmly announced that my son would suffer severe brain-injuries for the rest of his life before coolly ushering us out of his office. We travelled home and looked up the meaning of the various labels he used in a medical dictionary! Nowadays, you would probably use the internet, but should you have to? Why should you be treated in this way?

When one considers the devastation, which the deliverance of news such as this can cause, one would imagine parents would be informed as tactfully and discretely as possible. One would also hope parents would be informed together, in warm, hospitable surroundings and be given as much time as necessary to absorb the magnitude and complexity of the information. This certainly does not seem to be the experience of many parents and to this date, there are many research articles in the literature, which demonstrate the disclosure of such diagnosis by medical professionals to be an ongoing difficulty.

One mother related a particularly bad example of how the 'news' was broken to her when she explained to me;

"We sat facing the doctor, who was emotionless and cold. He just said, "I'm sorry but David's got autism." – That was it."

Another mother described to me the way in which she was informed about her little girl's problems.

"He is autistic and will be for the rest of his life." With those words ringing in our ears, we were ushered out of the office.

Out of the myriad of parents to whom I have spoken down the past few years, only once have I ever met parents who were satisfied with the way in which this situation was handled. These assertions are supported by research, which demonstrates that only twenty percent of parents are satisfied with the way in which the news of their children's problems was broken to them. It is possible to do this task properly though, as is proven by the study carried out by Nursey, (1991). She achieved a one hundred percent satisfaction rate in an experimental group of parents, who were informed of their children's' problems together, in private and in a direct, honest and sympathetic way and who were given immediate and easy access to support services.

A system, which leaves a delicate task as breaking such news to people who are trained to be calculating clinicians, when precisely the opposite traits are required, is seriously flawed. Would it not be more appropriate in such a situation to have someone available who is not a 'cool, calculating clinician,' someone who *is* trained to deal with people on a more human level, who could deliver the news in a more tactful manner? The professional resource is certainly available to do this and utilising it would save parents from having to deal with the trauma of receiving such shocking news, delivered in such a dreadfully incompetent manner. Psychologists and other counselling professionals abound and are qualified to deal with every conceivable stressful situation. This professional resource should be tapped, with experienced professionals on hand to talk parents through this dreadfully sensitive situation and beyond. They should act as the parent's 'advocate' with the medical profession and as such would be in a position to provide first hand information and support to new parents. They would be able to guide the family through the emotional minefield, which is to come.

Bringing such a system into the 'front line' instead of tolerating a situation where overwrought, confused parents have to find information and access to support services on their own would also eradicate other difficulties.

- Many parents who are dissatisfied with the explanations and information concerning their child's problems, given to them by the doctor (and there are many!), would be satisfied.
- Input from various disciplines could be more accurately tailored to

parental need.

- Parents would be more expertly guided through the trauma of coming to accept their child's problems, than they ever could be by someone whose training was purely clinical.

Will the medical profession heed these words? No! I predict that parents will still be voicing the same complaints in twenty years time.

What is autism?

So the diagnostic meeting with your doctor has left you in a state of shock? Hardly surprising! Shall we try to remedy your natural state of confusion with a logical look at the problem you are facing? OK! There are three major categories of autistic type labels which may be attached to your child.they are ***autism, asperger's syndrome and pervasive developmental disorder.(PDD).*** There are also disorder's such as Rett's syndrome which cause autistic type symptoms and children with cerebral palsy can also experience some of these symptoms, so we are looking at a wide range of children.

First let's look at classical autism, (including PDD) how would we recognise it? Well, autism was first recognised in the mid 1940's by a psychiatrist called Leo Kanner. He described a group of children, whom he was treating, who presented with some very unusual symptoms such as; - atypical social development, irregular development of communication and language, and recurring and obsessional conduct with aversion to novelty and refusal to accept change. His first thoughts were that they were suffering some sort of childhood psychiatric disorder.

At around the same time that Kanner was grappling with the problems of these children, a German scientist, Hans Asperger was caring for a group of children whose behaviour also seemed irregular. Asperger suggested that these children were suffering from what he termed '***autistic psychopathy***.' These children experienced remarkably similar symptoms to the children described by Kanner, with a single exception. – Their language development was normal!

There is still an ongoing debate as to whether autism and Asperger's syndrome are separable conditions, or whether Asperger's syndrome is merely a mild form of autism.

What is the cause of autism?

In the 1960s and 1970s there arose a theory that autism was caused by abnormal family relationships. This led on to the 'refrigerator mother' theory, which claimed that autism in the child was caused by cold, emotionless mothers! (Bettleheim, 1967). However the weight of evidence quickly put this theory to bed as evidence was found to support the idea that the real cause was to be found in abnormalities in the brain. This evidence was quickly followed by findings, which clearly demonstrated that the EEGs of autistic children were, in many cases, atypical and the fact that a large proportion of autistic children also suffered from epilepsy.

From this time, autism has been looked upon as a disorder, which develops as a consequence of abnormal brain development. Recently, evidence has shown that in some cases, the abnormal brain development may be caused by specific genes. (Collins et al, 2006).

However, we should not forget that genes can only express themselves if the appropriate environmental conditions exist for them to do so, (in the same way as genes which are responsible for alcoholism can only express themselves in a cultural situation where alcohol is available), and so, we should not rule out additional, environmental causes for autism.

We should not forget that autism can also be caused by brain-injury, that an insult to the brain can produce the same effects as can abnormal development of the brain, which may have been caused by genetic and other environmental factors. *(there are many within the medical fraternity, particularly those who are the subjects of lawsuits for causing brain-injury, who might understandably want you to think that the cause of autism is totally the responsibility of genes! They can then diagnose a child who has suffered brain injury due to medical negligence, with autism, - a genetic disorder. Anyone wonder why we have a sharp rise in the cases of autism?)* I have seen too many children who have suffered oxygen starvation at birth, who have gone on to display symptoms of autism or Asperger's syndrome. So, it is my view that autism can also be caused by brain-injury..

The brain is very delicate. Muscles can briefly utilise energy in the absence of oxygen, - the brain cannot. The brain is totally dependent upon its oxygen supply, which it obtains from the blood supply to the brain. A disruption in the blood supply to the brain of just 1 second, will see all the available oxygen in the environment consumed. A 6-second disruption produces unconsciousness. Within minutes, permanent damage is taking place. (Carlson, 2007).

So, whether for some children, autism is the result of some genetically inspired faulty wiring pattern in the brain, or whether for others it has an environmental cause, it all boils down to one thing. If a brain is not functioning normally, for whatever reason, it is injured. The cause of autism therefore needs no complicated definition, it is simple and easy to understand. The cause of autism is **brain–injury**.*(by whatever route)*. It is that simple! The important and simple thing to remember is that autism is caused by brain–injury; in fact, autism is an *expression* of brain–injury.

What do you mean by an *expression* of brain-injury?

When I speak of autism being an expression of brain–injury I am leading you down a path, which will hopefully allow you to see your child's problems in a new light. Let us take this a little deeper and say that autism is also a *symptom* (or set of symptoms), of brain-injury; - it is the way in which your child's brain–injury *expresses* itself. There are other expressions of brain–injury, some of which are used interchangeably with autism! I put the idea to you that autism is a symptom of brain–injury in exactly the same way that a cough is a symptom of a chest infection. The only reason for the existence of the term autism is that it, itself is a convenient marker for specific groups of sub-symptoms; - yes sub-symptoms of brain–injury! These symptoms might include epilepsy, poor visual, auditory and tactile development, impaired motor function, poor language development, difficulties of socialisation, learning difficulties, or other symptoms.

That is all very well, but are you not just 'nit -picking' over terminology?

No! It is important that this situation is both clarified and simplified. I have misgivings over the continued use of the terms 'autism' and 'Asperger's syndrome,' (as I have misgivings over the continued use of many other labels such as 'cerebral palsy,' for example). These are terms, which merely describe symptoms. I believe this is misleading and overly-complex. It is much easier to understand the concept and the possible ramifications of brain-injury, than it is to understand the concept of autism. I feel it is important that you should be aware of the problems around terminology and how it has encouraged the development of symptomatic treatments. Consequently, where I use those terms within this text, it is merely to make a point in reference to a particular expression of brain-injury.

In most other areas of medicine, it is the cause of an individual's problems, which receives treatment, yet in the case of brain-injury, this invariably does not occur. If your child is suffering from allergies and he will not stop sneezing, does the doctor place a peg on his nose and claim that because he has stopped him sneezing, he has cured the problem? No, of course he does not. He will (hopefully!) send your child for tests to determine the *cause* of the problem and then, having determined the cause, he will set about treating it.

I believe that in the case of brain-injury the medical world has fallen into the trap of merely trying to offset the worst effects of the symptoms. In some cases, as in the case of epilepsy, controlling the symptoms is a necessary step, but this should be done alongside attempts to treat the cause of those symptoms, as in some cases the focus on treating symptoms can have the effect of making them worse.

Reaction and Adjustment

It is recognised by most parents and professionals with whom I have spoken, that there are definite reaction stages, through which most parents pass after the discovery of their child's brain–injuries. I have no doubt you will, or already have experienced the same process, so here we go!

The initial reaction, as it was in my own case, after the diagnosis of my son, is usually one of shock, followed by reactions such as disbelief, anger, confusion and even guilt. In my research down the years, I have consulted many parents concerning their feelings during this initial period. All of them without exception confirmed reactions similar to those described above.

This period of confused emotion is usually closely followed by grief, as parents mourn the loss of the child, which 'should have been.' This may seem a strange concept to the outsider, but the child, which has been nurtured for nine months inside his mother's womb, is simply no more. Instead he has been replaced by another child, - one with terrible problems. To many parents, it is as though a child had died. One mother I know pointed this out to me very succinctly;

"The little boy I had carried for nine months had gone. I felt I had come to know him and then he had died. As he grew up, he had been replaced by a little boy who appeared to not want to know me. I was left looking after a stranger."

These feelings are sometimes very strong and can cause serious problems for families, who, because of the social isolation of many children with autism, find it difficult to relate to and bond with their developing child.. These thoughts and emotions are usually underrated by medical professionals and in the vast majority of cases little support or counselling seems to be offered. Families are left to their own devices as they desperately attempt to come to terms with the enormity of their child's problems.

It is now that parents seem to branch into two readily identifiable categories as follows.

(1)The conformists. These parents appear to fall into three further subcategories.

- Parents who believe their child needs love and protection and anything, which goes beyond this, is not even to be considered. As one mother stated. *"We think any kind of stimulation would be cruel, we don't want anything like that."* I place no criticism or judgement on this point of view, I understand it completely.

- Parents who appear to have been programmed into believing that the doctor is always right and consequently are only open to this single channel of information. The problem is, this channel of information can be as unreliable as any other! I recall when my wife and I were considering taking our son for treatment to the Peto Institute in Hungary. A doctor visiting our home, whose field was child development, tried in the most intolerable way to discourage my wife from taking this entirely reasonable course of action by saying; - *"I would consider Hungary very carefully if I were you, after all you will be away from home for several months. Whilst you are away, your husband could quite easily move someone else into your home!"*

- The third subgroup of the conformists are the 'reluctant accepters.' These are the parents who are interested in exploring every possibility, which may be utilised to help their children, but are constrained by some circumstance, which is outside their control from doing so.

(2).The non-conformists. – These parents are the ones who are likely to pursue other possibilities of treatment for their children and they appear to fall into two further subgroups.

- Parents who have not completely adjusted to their child's problems and are determined, (sometimes unrealistically), to disprove all the experts who have written off their child. These parents unfortunately have not made it to the final stages of adjustment to their child's problems. Again, I do not criticise them. *(for many years I fell into this group)*

- Parents who have adjusted well to their child and his problems and despite all of the adverse medical 'opinion,' see no harm in exploring all possibilities, which may help their child. (And there will be adverse opinion, simply because you will be exploring strategies for treatment, which are outside the realms of medicine: - many medical people will perceive this as some sort of threat.

Ok, so you have probably fallen into one of the above categories and you have decided which course of action, (or inaction) you are going to embark upon. This is entirely your choice and no one has the right to be critical of you whatever that choice may be.

Whatever your decision, you need to know more about how autism is going to affect your child, so let us begin by examining the impact that this type of brain-injury can have on your child's sensory system

Problems of sensory perception

Sensory problems are common amongst children with autism, the fact that they are often overlooked is sad because they are the cause of so many more 'knock–on' effects. Many children with autism experience a great deal of difficulty in modulating sensory experience, which can make the environment in which they live a very confusing, threatening and frightening place. Let us take a quick look at the sensory system and the types of difficulties autistic children face.

Although the sensory system is very complex and its correct development is vital, this book is only able to provide a brief guide. I shall therefore highlight the major problems, which children who suffer this type of brain-injury face in the developmental areas of vision, hearing and touch.

We take in information about our environment through our senses. This is something, which we cannot help but do. The amount of sensory information our brains are processing at any one moment is phenomenal. As I sit here typing, I am aware of several sensory stimuli. Visually I see the computer keyboard, with the letters printed on the keys; I can see my hands, the desk, the computer screen and more. In my peripheral vision, I am aware of the window, my dog, the sofa and other items in the room. Auditorially (hearing), I can hear the kettle beginning to boil, I hear my fingers tapping on the keys of the keyboard and I hear my cat mewing. In terms of touch (tactility), I can feel the keys of the keyboard; feel the wooden floor beneath my feet etc. These are merely the things of which I am aware and these sensory stimuli are all being processed simultaneously, in a fraction of a second.

As an example of this processing, consider the complexity of my typing this text, which hopefully you are reading with enjoyment! The front part of my brain (the ***frontal cortex***) is sending out messages to areas of the motor parts of my brain (the ***motor cortex***), which control my hands, instructing them which keys on the keyboard I need to hit next in order for the written words on this page to make sense. The motor cortex then instructs the hands to move in order to hit those keys. Parts of the brain known as the ***basal ganglia*** and ***cerebellum*** then become involved in order to attempt to execute the necessary movements of the hands in a fluent and accurate manner. When the movements have been executed, feedback signals are then sent back to the frontal cortex, via the 'cerebellum' and 'basal ganglia' to inform it how successful the hands were at hitting the correct keys on the keyboard and whether the movements were accurate and fluent. If necessary, the frontal cortex then issues new instructions, to correct any errors.

In a healthy, uninjured brain, this grossly oversimplified description of events all takes place within a fraction of a second whilst the brain simultaneously takes care of many other complex tasks. It is a phenomenal feat.   Also, consider how the brain decodes the various sounds we call language and how it regulates its own attention. Imagine you are sitting in your lounge holding a conversation with a visiting friend. There you are, happily chatting away; - you are attending to your friend's voice so that your auditory system is able to process the constant stream of noise, which we call speech. Your brain is able to take this constant stream of sound, break it down into recognisable chunks and attribute meaning to it so that you understand what it is your friend is saying. At the same time, your brain is *tuning out* extraneous sounds in the background, such as traffic passing outside your window, so that you are able to focus on the task at hand. Your brain accomplishes all of this and much more, (This again is actually a gross oversimplification), with the minimum of effort, without you even being conscious of the processes involved.

Now consider a brain, which is not healthy; - a brain, which has suffered injury and try to imagine the chaos that might ensue for a child whose sensory processing system has been impaired. Imagine this child's ability to 'tune out' noises, which he does not wish to pay attention to, has been impaired. What havoc would that child experience?

It was Delacato in the seventies who first claimed that children suffer distortions of sensory processing, separating them into the categories of 'hyper-sensory,' hypo-sensory' and white noise. I do not feel that this classification discriminates finely enough. I have managed to identify five categories of sensory difficulties, which children display, (other researchers may find more!) and see these as symptoms of a malfunctioning 'tuning mechanism' in the brain. This 'tuning' mechanism is the structure which enables us to 'tune out' background interference when we wish selectively to attend to something in particular; it also enables us to 'tune in' to another stimulus when we are attending to something completely different. It is the same mechanism of the brain, which allows us to listen to what our friend is saying to us, even when we are standing in the midst of heavy traffic on a busy road. It is this mechanism that allows us, even though we are in conversation in a crowded room, to hear our name being spoken by someone else across that room. It is this mechanism, which allows a mother to sleep though various loud, night-time noises such as her husband snoring, or an aeroplane passing overhead and yet the instant her new baby stirs, she is woken. It is a remarkable feature of the human brain and it is the responsibility of three structures operating cooperatively; - these are the ***ascending reticular activating formation***, the ***thalamus*** and the ***limbic system.***

Having made such a bold claim, allow me to furnish you with the evidence to support it. The three structures just mentioned receive sensory information from the sense organs and relay the information to specific areas of the cortex. The thalamus in particular is responsible for controlling the general excitability of the cortex (whether that excitability tunes the cortex up to be overexcited, tunes it down to be under excited, or tunes it inwardly to selectively attend to it's own internal sensory world.) (Carlson, 2007). The performance of these neurological structures, or in the case of our children, their distorted performance seems to be at the root of the sensory problems our children face and yes, their performance ***CAN*** be influenced, - they can be re-tuned.

I would label the five categories of sensory distortion, which I have witnessed in children with autism as follows: -

1. **Sensory over-amplification**. The particular sensory modality, (vision, hearing, touch, etc) can become oversensitive to stimuli from the environment. It is my belief that in this case, the thalamus, limbic system and reticular formation, which are acting as the brain's 'tuning system' are malfunctioning and are not effectively regulating the level of incoming sensory stimuli. Indeed, in this case they would appear to be acting to over-excite the cortex, which would have the effect of amplifying the sensory stimuli. This could possibly cause the child to overreact, or to withdraw into himself as a defensive strategy.

2. **Sensory under-amplification.** The particular sensory modality can appear to become under sensitive to incoming stimuli from the environment. In this case, I believe the thalamus and other two brain structures, acting as the tuning system, are acting to

under-excite the cortex, which is having the effect of appearing to dampen down incoming sensory stimuli. This could influence the child to act as though he cannot see, hear or feel; - he may be deficient in one or more sensory modalities.

3. **Internally focussed sensory tuning.** In this case, the particular sensory modality appears to be 'inwardly tuned.'. In this case the three brain structures, acting as the brain's tuning system are exciting the cortex to attend to sensory information of the sensory system's own making, or from within the child's own body. Therefore, the child may have difficulties perceiving the 'outside' sensory world through this haze of internal stimulation.

4. **Wide spectrum tuning.** In this case, the three neurological structures are exciting the cortex and attempting to tune its attention to many incoming stimuli simultaneously. They seem unable to filter out background noise, sights, etc in order to allow the child to focus on one aspect of the environment. For this child, the world is absolute chaos and again, he often withdraws into himself.

5. **Narrow spectrum tuning.** In this case, the neurological structures are only exciting the cortex selectively, allowing the cortex to attend to limited, isolated sensory stimuli. This child may often seem 'over-focussed' on one particular aspect of his environment. He can for instance, become intensely interested with a spinning top or the particular features of one toy and will not play with anything else, to the point of seeming obsession. For this child, it appears his sensory tuning system is focussed too narrowly and he cannot spread his attention to incorporate several features of his environment simultaneously.

As we can see, brain-injury interferes with the maturation and development of the sensory system in a number of ways. Quite simply, it will have stopped, slowed or distorted the child's development. Now let us examine the major areas of the sensory system and explore in a little more detail how brain-injury may have affected your child.

Vision

Many people regard vision as their most important sense. I would agree and the neurology seems to support this view. 25% of the human cortex is devoted to processing visual information. With that huge amount of cortex devoted to just one function, the chances are that any diffuse brain-injury will knock out a significant portion of brain cells dedicated to vision. Is it surprising therefore, that so many brain-injured children experience visual difficulties? Let us take a look at some ways in which those visual difficulties can express themselves.

Wide spectrum tuning

Imagine a situation where everything within your visual field is competing equally for your attention. In this situation, you would not have the ability to 'tune out' some elements of your visual field and selectively attend to one or two elements alone. Your brain would be trying to process everything you could see *at the same time!* The result is chaos for the children who suffer this problem, causing anxiety and high stress. The problem was aptly described by Bruno (2006), who endured brain-injury through a car accident. She described how brain-injury propelled her into a world of psychological perplexity, double vision and incapacitating visual and auditory oversensitivity

Many parents report that the visual world is just too much for their children, indeed children have themselves reported a situation where they are unable to focus on a single visual stimulus. Indeed, it seems that apart from inappropriate activity from the brain's tuning mechanism, damage to parts of the brain's **parietal lobes** can lead to this type of visual difficulty. A child may appear to be competently scanning his visual environment, but cannot attend to particular visual features of that environment! (Carlson, 2007)

The child who suffers this problem will only relax if placed in an under-stimulating, darkened environment. He is only truly at peace when he is asleep. He rarely makes eye-contact, because he has difficulty attending to the single visual stimulus of the eyes, which are competing with other visual stimuli in the environment (does this ring any bells for parents of children diagnosed with autism?). Everything is competing for his visual attention simultaneously, so he finds it impossible to focus on any one person or object.

As a consequence of his inability to cope with his visual world, the child with this problem can, in a desperate measure to protect his immature, overwhelmed sensory system, 'withdraw' into himself and appear to be difficult to reach.

Narrow spectrum tuning

In this situation, the child appears not to be aware of most of his visual environment, singling out one object and almost completely focussing his attention on it. This 'over-focussed' attention can appear to be obsessive behaviour to the outsider. This child plays with one toy and one toy only because he is focussed upon specific features of it. He may be fascinated with movement, such as a spinning top, or wheels moving and will spend hours just looking at this. Rizzo & Robin, (1990), describe this situation perfectly. Apart from a malfunctioning neurological tuning mechanism, injuries to the ***parietal lobes*** of both hemispheres of the brain can also create a situation whereby individuals can only pay visual attention to one object at a time. This is known as ***Balint's syndrome*** and it is possible that some of our children, who experience 'narrow spectrum tuning' difficulties, may have injuries here. However, there is also a convincing developmental explanation for 'narrow spectrum tuning.' Young babies have difficulty in shifting their attention, - this is well known and is a developmental phase. Infants who are less than four months of age will sometimes stare at an attractive object, being unable to shift their gaze. Occasionally this inability to shift their visual attention will make them cry out in distress (Johnson et al, 1991). It could very well be that children who have 'narrow spectrum tuning' difficulties, never emerge from this phase of visual development.

Over-amplification

This again is a type of visual oversensitivity whereby the sensory tuning system of the child is acting to amplify the visual information, which the eyes are taking in. The child who suffers visual over-amplification problems, is the child who hates bright lights. He particularly dislikes sunny days and hates anything moving near to his eyes.

He may not concentrate on anything at all with his central vision, preferring to view things from the less threatening position of his peripheral vision. We, as healthy individuals are allowed a small insight into the way this child feels when we have a migraine attack and our vision becomes sensitive to bright lights. What is occurring both in migraine sufferers and with our brain-injured children is that specific inhibitory systems of the tuning system within the brain are not activated sufficiently, resulting in overstimulation in the visual cortex. (Mulleners et al, 2001).

Unfortunately, this is the visual world, which this child inhabits 24/7. This child will not make eye contact, but for different reasons to the child with wide spectrum tuning difficulties; - this child literally finds eye contact to be very threatening and will avoid the situation at all costs.

Under-amplification

In this situation, the sensory tuning system of the brain is simply not exciting the cortex sufficiently and so it is unable to process incoming sensory information. These children are sun–worshippers; they find bright lights and visually attractive displays to be fascinating. Children with this problem, if their motor control allows, can often be found waving their hands in front of their eyes in an attempt to self stimulate their visual system.

Children showing under-amplification problems, like those displaying over-amplification difficulties can seem hard to reach, but for the opposite reason; - they are simply unaware of much of the visual world around them.

Internal tuning

In this phenomenon the visual system is tuned inwardly to visual phenomena, which it itself is creating. Again, it is possible to relate this to what happens to certain individuals who suffer from migraine. Many migraine sufferers, (including myself!), experience a 'visual display' prior to an attack, where all sorts of shapes and colours occlude the vision. Similarly, the visual system of some brain–injured children is capable of producing this effect. These children appear preoccupied with looking at something, which you cannot determine! They appear to be staring into the mid–distance and it is immensely difficult to break their concentration. As early as 1956, Beck and Guthrie were describing the internally generated visual phenomena experienced by individuals who had suffered brain-injuries, one describing seeing different coloured orbs in their visual field, floating up and down. (p. 6). Is it any wonder that some of our children are fixated upon this self–generated visual world?

Other problems of visual development.

The development of vision and the child's ability to use his visual skills in a meaningful way may be, as I have just described, distorted by brain-injury. Visual development therefore will most likely be stopped, or slowed to snail's pace. Injury may interfere with the smooth operation of the visual pathways in the brain, or cause direct injury to part of the brain known as the ***occipital cortex***, which is the processing centre for vision. Injuries such as these can take a terrible toll. They can take the visual ability of the child back to pre-birth stages, in some cases creating a ***neurological blindness***. - This is a situation in which there is nothing at all wrong with the eyes, they are working as they should, but because of damage to the primary visual cortex and the fact that essential parts of the neural networks, which support visual ability have been damaged, the brain is simply unable to process what the eye can see.

I remember one little boy, who I was requested to see, who was very unresponsive in visual terms. He did possess a pupillary light reflex, (his pupils dilated when in the dark and constricted when in the light). His doctors, who had informed the parents that he was probably cortically blind, were surprisingly, not doing anything to try to remedy the situation. It took the parents two years of patiently stimulating their son's visual development under my direction, to bring his vision to a level where he could visually track an object across a room and visually explore his environment. The most moving moment however, was the first time he looked his mother in the eye and smiled. From there, I instituted a reading programme and he made incredible progress.

Other visual problems, experienced by brain-injured children include visual field problems. - Each hemisphere of the brain is responsible for processing visual information from the opposite half of the visual field, so an injury to part of the occipital cortex in the right hemisphere of the brain can cause a visual deficit in the left visual field and injury to part of the left occipital cortex can cause deficits to the right visual field.

Another phenomenon, which can occur due to injury to the occipital cortex, is that the child may not notice movement within his visual environment. He may pay good visual attention to most aspects of his visual environment and yet fail to detect the sudden movement of an object close by.

Deficits in the ability to perceive colour (***cerebral achromatopsia***), may also be experienced due to brain-injury. Interestingly, this problem may occasionally be experienced in only one half of the visual field, if the injury to this part of the occipital cortex is only in one hemisphere. Children with this type of injury in both hemispheres report their vision as being in black and white. (Heywood & Kentridge, 2003).

The path of visual development

When a child is born, his vision is already at a relatively sophisticated level. He can see quite well; his vision is a little blurry and he cannot see as far or as clearly as you or I, but essentially, he can see reasonably well. He does have difficulty switching his focus from one point to another point, which is at a different distance. He is however able to scan his visual field, although at this point his eye movements are slow and disjointed.

By the time he is one month of age however, his eye movements are smooth and he is able to scan his visual field more effectively.

You may notice that a young baby may appear to have very big eyes in relation to the size of his head; - There is a very good reason for this. The eyes of a young baby are forming a massive number of complex attachments with the brain. If the eyes grew substantially after making those attachments, new nerve fibres, (***axons***) would have to be grown in order to make new attachments between the larger eyes and the brain. This would mean that the brain would have continually to reorganise its attachments as the eyes grew. Hence the eyes come 'ready made,' full size! (Kalat, 2001).

At two months of age, baby is developing ***pattern discrimination*** and ***contrast sensitivity***. At this point, he prefers less complex patterns. (Such as preferring to look at a checkerboard with large squares, rather than one with small squares).

Infants are very attracted to looking at high-contrast edges and patterns. Large black and white patterns offer the maximum achievable contrast to the eye and consequently are the most noticeable and eye-catching to babies.

By three months of age, baby is able to focus as well as an adult and at four months his pattern discrimination has developed to the point where he prefers to look at more complex patterns and is becoming interested in the internal detail within a shape.

At six to seven months of age, when he is starting to crawl, he is beginning to use his two eyes together and is developing an appreciation of the third dimension and depth perception (***stereopsis***). There is evidence that the specific neural networks, which are essential for the development of stereopsis, will not develop unless baby is provided the opportunity to scrutinise objects with both eyes. If baby's eyes are not given practice in moving together properly, he may never develop stereoscopic vision, even if the eye movements are later rectified by surgery on the eye muscles (Banks et al, 1975).

There is also some evidence that it is ***crawling***, which helps the development of depth perception by helping to mature the relevant areas of the brain and by affording the child, through movement, the opportunity he needs to use both eyes together properly.(Berk, 1997).

Vision continues to develop throughout the preschool years. It is essential that it does so, in order that there are continued improvements in eye/hand coordination and depth perception. There are many exercises, which can be carried out with brain-injured children to try to achieve these objectives.

One of the most important and enjoyable exercises to carry out with young children is to read to them. This encourages the development of robust ***visualization proficiency*** as they "picture" the story in their minds. Just because a child has suffered brain-injury is no reason to deprive him of this enjoyable activity. Although his vision, or hearing may be impaired to some degree, one never knows how much of the message is actually striking home; - so read to him!

By school age, the child's visual acuity, (the level of fine discrimination of detail, which his vision will permit), is equal to that of an adult.

Auditory development and language comprehension

Auditory development, as the term suggests, is vital for the development of our ability to hear and consequently the development of the ability to understand and produce language. Without hearing, language development is severely affected. With distorted auditory development, language development is itself distorted. We will now review how distorted auditory development can affect the child with brain-injury.

Wide spectrum tuning

Evidence has accrued over many years, which suggests that the brain extensively analyses sounds to which we are apparently not paying attention. However, having analysed these sounds within our environment, the brain decides that the vast majority of sound is unimportant, so it receives no further processing and no further attentional resources are diverted toward it. In other words, we actively 'filter' out sounds, which our attentional systems consider to be unimportant. So what happens when a child who has suffered autism type brain-injuries loses the ability to focus his hearing and to filter out sounds efficiently?

Auditory wide spectrum tuning is one of the most difficult problems a child and a family can face. Children with this problem experience difficulties in devoting attentional resources to sounds to which they need to pay attention and conversely have difficulties in 'filtering out' sounds to which they do not need to pay attention. Simply, the brain is attempting to devote attentional resources to everything in the auditory environment. To understand this, again place yourself seated in your lounge, having a conversation with a friend. You have no difficulty in attending to your friend and what she has to say. A car passes by the window whilst you are talking and you completely ignore it, you may not even be aware that it has happened; you have successfully 'tuned out' the sound of the car and selectively paid attention to what your friend was saying.

In sharp contrast however, the attentional and filtering system of the child with brain-injury may not be able to succeed in accomplishing this sophisticated task. Because of the breakdown in his ability to filter out non-important sounds, his system attempts to attend to and process everything simultaneously! – All noise in the environment is concurrently competing for his attention. It must be a world of unimaginable chaos!

Children who suffer this type of difficulty can understandably present as being hyper-anxious and agitated. The child may withdraw completely into himself as a necessary defensive strategy in order to try to escape a chaotic world. Many of these problems have the 'knock on' effect of slowing or completely stopping the language development of a child, as he cannot separate the speech sounds of the human voice from all other background noise.

I will never forget the state of one little chap who's parents brought him to see me. They wheeled him into my office and I knew instantly what his major problem was; - he was experiencing sensory overload. His muscle tone was rigid, his breathing was strained and anxious and he was obviously *very* aware (too aware) of everything which was happening around him. I privately wondered how the child managed to survive each day. I informed the parents that in my opinion, the 'tuning system' in their son's brain was malfunctioning and in particular his brain was attempting to process too much of the auditory stimuli in the environment. To prove my point, I made sure the general noise level was reduced and asked that we should all lower our speaking voice to a whisper. Within a few minutes, the child was much calmer. I recommended a programme of sound reduction, auditory training and vestibular stimulation, which had the effect of further calming the child.

Narrow spectrum tuning

Human beings have an amazing capability to focus auditory concentration, (this is particularly the case concerning spoken language), to the utterance of one specific speaker who may be talking within the noise created by a loud group of people. Recordings of brain activity with positron emission topography (PET), advocate that a complicated set of connections, which comprises several brain regions, (auditory, frontal and parietal areas of the cortex), are involved in this type of selective attention (Tervaniemi et al, 2000).

The child, who suffers narrow spectrum tuning difficulties, over-focuses his attention and has difficulty widening that attention to incorporate other salient features of his auditory environment. Outwardly, his behaviour can easily cause him to be confused with the child who suffers 'under-amplification' difficulties. His tuning system only allows him to attend to one stimulus at a time and so he can give the impression of being under-sensitive to all other sounds in his environment. This child, in his attention to sound, can appear to obsess on one particular feature of the environment to the exclusion of all else.

Haist et al, (2005), in their research on specific types of brain-injury and its effect upon the individual's attention, point to the fact that certain patterns of injury have a tendency to produce an attentional system which over-focuses. These findings corroborate my own experience in dealing with children who face these difficulties.

Under-amplification

Children, who experience auditory under-amplification difficulties, love loud noises. I knew one child who, if an air horn were used right next to him, would react with pure delight. (Please do not try this until you have ensured that your child definitely suffers under-amplification problems by having him professionally evaluated). Children who experience this problem can become very frustrated as they desperately attempt to make as much noise as possible in order to self–stimulate their auditory system. In the case of children such as this, the brain's sensory tuning system (the thalamus, limbic system and reticular formation), may not be exciting the cortex sufficiently to allow it to process sensory stimulation from the environment. It is also the case, as it is with their oversensitive counterparts, that some of these children can be selectively undersensitive to particular frequencies. This can be reflected in their language development, as they will have difficulty in receiving, processing and consequently producing specific speech sounds.

Over-amplification

The world is simply too loud and oppressive a place for children who experience this problem. Every sound is amplified. Children with this problem can appear not to be hearing anything, as they 'close down' the threat of the auditory world. They simply cannot deal with a world, which is too loud and threatening.

Children who suffer auditory over-amplification difficulties react badly to hearing the frequencies to which they are sensitive and they fare badly in acoustically unfriendly places such as bathrooms. I knew one child who would scream hysterically every time his parents used the vacuum cleaner, to the point where the parents were forced to take him out of the house when it needed to be used. Other children I have known have hated high-pitched noises and demonstrated some quite unusual reactions to them. I know one little chap who 'giggles' every time he hears a whistling sound, which sounds amusing, but this is no 'giggle' of enjoyment, - if you looked into his eyes you would know instantly that it is false and that the noise is actually causing him pain.

Some recent research has also linked this type of sensory over-stimulation difficulty to an under-production of a chemical called **5-hydroxytryptamine**, a precursor to **serotonin**, (Marriage, 1995). This would support my observations, which connect children's sensory over-sensitivities with the development of a poor sleeping pattern, (correct serotonin levels are also necessary for the smooth operation of the sleep / wake cycle).

One little girl who I was privileged enough to see, clearly suffered problems of auditory over-amplification. To the untrained eye she seemed passive enough, sitting in her chair apparently at peace, (she was actually 'closed down,' trying to keep a distance from the threatening world around her). There were however little tell-tale signs that she was not at peace and as I observed her I formed the conclusion that she may in fact be over-amplifying. To test out this hypothesis I took a tuning fork and gently tapped it on the edge of the desk. Her reaction was immediate and startling: - She screamed in pain! My lesson was well and truly learned and I have never repeated the tuning fork test!

Internal tuning

Ever hear the old saying, 'I've got bells in my ears?' Anyone who has experienced this phenomenon for a while will comprehend a little of what these children have to endure; - people who suffer from 'tinnitus' will certainly understand. This child cannot focus on the outside world, because his tuning system is forcing the cerebral cortex to attend to the noise, which is being produced by his own sensory system. He can appear to be barely aware of what is occurring around him. He may be intently listening to his own internal sounds, such as the pulse of blood being pumped through his ears, or his sensory system may be producing its own auditory *white noise* on which he is fixated. Many authors term this phenomenon as an ***auditory aura,*** which emanates from injury to or over-excitation of the ***primary auditory cortex.*** Hughlings – Jackson first connected the auditory aura to a specific type of epilepsy known as ***Jacksonian*** epilepsy.

Hughlings-Jackson described three cases of the auditory aura, which he integrated with his concept of the "dreamy state" (now known as mesial temporal epilepsy). In doing so, he described the current concept of elementary and complex auditory auras. Hughlings-Jackson also associated the auditory aura with seizures that began in the "primary auditory cortex."

Children with this type of problem almost seem to be in a 'dream state,' locked into a world from which they seemingly cannot be retrieved. Their concentration appears to be on something, which you simply cannot hear, which of course it is!

The path of auditory development.

Baby has been hearing its mother's voice in particular, from inside the womb, for a while. Is it any wonder then, that at birth baby is already demonstrating a preference for and responsiveness to the human voice? What is very noticeable at birth, is how easily baby startles and is frightened by loud noises. This response is inbuilt and is a vital response to what could constitute a threat: - This ***startle reflex*** is designed to alert his carer that he may be in danger and to elicit their protection and comfort.

At one - month of age, baby is already demonstrating sophisticated abilities; - turning his head towards a sound, enjoying gentle music and showing preference for the voice of his parents amongst other voices.

At three months, baby is responding appropriately to tone of voice, (He will smile if spoken to gently and cry if spoken to sternly) and is beginning to demonstrate that he appreciates meaningful sounds. For instance, if he hears water running he will anticipate being bathed and will either cry or become excited depending upon whether he enjoys or dislikes bath-time.

At around four months, he begins to be able to reach in the direction of sounds and can show a preference between consonant and disconsonant music, (some children with damage to the ***left parietal lobe*** of the brain are completely unable to perceive music) (Peretz, Gordon & Bouchard, 1998: Zentner & Kagan, 1998).

At seven months, he begins to show an initial understanding of human language, recognising his own name and will begin to try to mimic sounds, which he regularly hears in his own environment.

At around nine months, he will begin to search for people when asked to. If he is asked, 'Where's Mummy?' he will look around for her.

By twelve months, he is clearly demonstrating a more sophisticated understanding of language, which may extend to as many as forty words of speech. He is also beginning to understand some phrases. He will now also respond to simple requests such as 'pick up the ball,'

By fourteen months of age; - about the age he has started to walk, he is demonstrating a further level of sophistication in his understanding of language, responding to requests such as 'bring the ball to Mummy.' He should also by now be aware of all the names of close family members.

By eighteen months, he is beginning to show understanding of sentences and displays immense enjoyment in listening to children's songs. He is also demonstrating an enjoyment of and sensitivity to children's rhyme, which is very important, as sensitivity to rhyme has been linked with future reading ability. (Bryant & Bradley, 1996).

By thirty months, he is showing clear enjoyment of bedtime stories and now understands language in a highly complex manner.

At five years of age the journey is largely completed, he now understands the basic vocabulary and grammatical structure of his language, although he will continue to add words to his mental lexicon for years to come.

Mental lexicon: - The mental lexicon is a structure or group of structures in the brain, where words and their meanings are stored.

As you have seen from the previous pages, brain-injury can have a catastrophic effect on the auditory system. In addition to its effect upon the sensitivity and tuning of a child's auditory capabilities, it can also have other developmental implications. In a similar fashion to visual development, brain-injury does not affect the mechanical apparatus involved with hearing, - the ear and associated mechanisms. What it can do is to cause neurological deafness. This means that although the ear is perfectly capable of receiving sounds, the brain does not interpret them. Again, in the case of auditory development, it is clear that it is an orderly developmental pathway. Later, I shall demonstrate how it is possible, with appropriate stimulation, to attempt to guide the child down all of these developmental pathways.

Tactile development

Quite simply, tactility is the sense of perceiving touch. If you were to be stroked gently, you would be receiving a gentle tactile experience. If a cat scratched you, you would be experiencing a harsh tactile experience. As with the other major areas of sensory experience, tactile reception can be disturbed by brain-injury.

Wide spectrum tuning

This problem is difficult to identify, but some people suffering autism have reported a situation where they are aware, in sensory terms of all parts of their body equally: - They are unable to focus on one sensory experience on one area of skin, as all areas, which are in contact with the environment, are simultaneously active. The child has difficulty in filtering out sensations to which he does not need to pay attention. The neurological structures, which over-excite the cortex, are working overtime with the consequence that it seems there are so many parts of his body, which are passing sensory messages through to his brain that he cannot selectively attend to an individual tactile experience. It is as though the message gets lost in the melee of the sensory overload, which his processing system has to deal with.

Narrow spectrum tuning

Children who experience this type of problem are only able to focus on one tactile stimulus at a time, so for instance, if you were to tickle their arm; they would be unaware of any other simultaneously occurring sensory experience. This can pose serious problems, as it can with children who suffer 'under-amplification' difficulties, because whilst focussed upon a single sensory stimulus, they are unaware of any other sensory stimulus in the environment, even potentially dangerous ones.

Many psychological representations of such narrow spectrum tuning presuppose that tactile attentional systems operate to filter out extraneous information in order to shield our limited-capacity processing systems from information overkill (Posner and Cohen 1984). In the case of our child who suffers narrow spectrum tuning difficulties, the filter is acting too efficiently, - its spotlight is too focussed.

It has been recognised that in primates and debatably in humans, the neural chains for tactile processing are ordered serially, with the main synaptic projections going from the ***thalamus*** to the ***cortex.*** Yet again, it seems we have the thalamus heavily implicated in the sensory distortions, which our children face.

Under-amplification

Children who are under-sensitive in tactile terms are the ones who 'bump' into things seemingly without hurting themselves. They can be covered in bruises and abrasions, which never seem to bother them. At the severe end of the scale, they are the children who can appear to 'self-abuse.' They may sometimes even go to the lengths of injuring themselves in order to try to feel some sensation. Parents have to take great care with children who suffer this difficulty as they may inadvertently cause themselves harm.

We have all seen or heard about children who can bite, scratch, cut and hurt themselves in multitude of ways and have thought, 'I wonder what makes them want to do that?' Well, have you ever considered what it may be like to have a sense of touch that is so dampened that you cannot really feel a great deal? You might begin to pinch yourself in order to feel some sensation. If that did not work, you might bite yourself in order to try to establish that there is some level of sensation. After many attempts at this and in utter frustration at still not being able to feel much sensation, you may become utterly frustrated and especially if you suffered other problems, which impeded your understanding, you may inadvertently begin to harm yourself, not understanding the implications of your actions.

This is the situation faced by many children who have under-amplification difficulties. The sensory system's tuning mechanism is simply not exciting the cortex sufficiently to allow it to process environmental sensory stimuli adequately.

Children with this difficulty can sometimes seem aggressive, making excessive physical contact with other people and touching other children in too forceful a manner. This child seeks out tactile experience and may seem 'clingy' in wanting to be constantly held and cuddled.

Over-amplification

Many children who suffer brain-injury, experience tactile difficulties. Children who are over-sensitive in this area hate to be touched, they can sometimes pull themselves away from all contact. Tactile experiences, which you or I might take for granted, can send them into paroxysms of fear. They may be sensitive to specific types of sensation. One young girl I know could not bear to have wool next to her skin, she would scream in apparent agony. It took her mother a while to discover what was causing the problem; - she removed the offending textile and the situation improved. I have also known a child who was oversensitive down one side of his body and under-sensitive down the other side! These children evade making bodily contact with other people and objects in the environment, leading to poor sensory understanding and social remoteness.

A child who is uneasy with touch may not feel secure and comforted by a parent's embrace, indeed he may avoid it like the plague.

Internal tuning

The concept of 'internal tuning' is a difficult one to grasp. Have you ever suffered from what is known as 'pins and needles,' (paraesthesia)? At some time, most of us have experienced this phenomenon. It is a sensation of uncomfortable tingling or prickling, usually felt in the hands or feet. The affected area is sometimes said to have 'fallen asleep'. A common cause is leaning or lying awkwardly on a limb, which either presses against the nerves or reduces the blood supply to the local area. Changing position soon restores normal feeling. Any numbness is quickly replaced by the tingling and prickling sensation, as the nerves start sending messages again to the brain and spinal cord. It is documented that in some cases, the 'pins and needles' phenomenon is caused by damage to the central nervous system. This state of affairs can be persistent and troublesome and can be caused by a large variety of sources, including brain-injury. This sensory activation can occur in any area of the body. Other types of paresthesias include feelings of cold, warmth, burning, itching, and skin crawling. Paresthesia occurs when an area of the body loses its normal sensation to touch. It may feel like a burning, pricking, tickling, or tingling sensation.

In some, but by no means all cases of brain-injury, this tactile phenomenon is a constant feature. The sensory system is creating its own 'interference,' which the tuning system is keying the cortex into, making it difficult if not impossible for the child to be aware of stimuli from outside his system; - from the outside world.

Problems of tactile development

The sense of touch not only helps to stimulate physical growth, it also plays an important part in emotional development. (Berk, 1997; Stepakoff, 2006). Could this be why children with the type of brain-injuries, which cause certain types of autism, are so emotionally distant? Tactile development is also an essential precursor to the development of motor abilities. In fact, the development of all 'output' abilities (language, fine motor development, mobility) are dependent upon and develop only after sensory abilities have developed. This leads us to a principle of treatment: - In any case of brain–injury, including what we know as 'autism.'- Sensory assessment and the development of a programme to enhance sensory development is a first priority.

One point to note with children, who in sensory terms are 'oversensitive,' whether it is visual, auditory or tactile oversensitivity is the chaos, which can ensue for the child,'s sleeping pattern. I realise how important this subject is to many of you who simply do not sleep, because your child does not sleep. – I was in this situation for sixteen years as my own brain–injured son was one of the 'night-hawks.' This will be discussed in more detail later, as 'sleep' has its own mini–section.

The path of tactile development

At birth, baby demonstrates quite a reaction to temperature and pain, especially to cold rather than warm. Consider what happens when you undress baby; - he quickly starts to cry. He is demonstrating a ***critical response*** to a distressing tactile experience. If you were accidentally to cause him pain, such as scratching his skin with your fingernail, his response would be a vigorous, high-pitched cry. (Look what happens when you take him for his vaccinations).

Baby's tactile discrimination soon becomes more sophisticated and by around two months of age, (as in all cases, some develop more quickly or more slowly than others), he is beginning to display more sophisticated discrimination between agreeable and disagreeable tactile stimuli. He will enjoy being stroked and tickled, but will show displeasure at stimuli, which are unpleasant, such as a scratch, or being in contact with a rough surface.

By six to seven months of age, his tactile awareness of depth is developing, as it also is in the areas of vision and motor development. He now begins to understand for instance, that something which appears to be flat, when placed on a surface, is not necessarily so! He becomes aware that it is possible to manipulate and explore the object from different angles. A great deal of 'exploratory mouthing' of objects occurs at this age.

In the following months, his tactile competence increases in sophistication even further, to the point where by the end of his first year, the child's tactile discrimination should have advanced significantly to the level where some familiar objects are beginning to be able to be distinguished merely by the use of touch. This ability improves and becomes more sophisticated and at approximately five years of age the child can discriminate most familiar objects merely by using his tactile senses.

Having highlighted the sensory difficulties, which brain-injury may pose for your child and explored the 'normal' developmental pathway which children take in each of the major areas of sensory development, it is now appropriate to move on to the developmental areas which I term the 'output functions,' – areas such as language development, fine motor development and mobility development.

Language Development

Not all children who have autism experience difficulties with language development. Children who have Asperger's syndrome do not experience delayed language development; indeed, they often display a superior command of spoken language. However, **all** children with autism or Asperger's syndrome suffer some level of **communication** impairment. The difference is in the distinction between **language** and **communication**. Communication does not have to be verbal; it can be nonverbal, body language for instance, or facial expression! Many children with autism have difficulty in producing and reading these nonverbal communicative signals.

It is **the** outstanding feat of the human brain that within eighteen months of birth, an uninjured human child, who has entered this world possessing no language, has an understanding of his native tongue and has begun to talk. That this leap into the use of symbolism is accomplished in what seems such an effortless manner makes it even more astounding.

The development of language and communicative abilities within a brain–injured child is literally the most vitally important matter to be addressed and is the key to overcoming other problems faced by the child. You may ask why? You may think that your child has so many pressing problems, both concerning his health and concerning his disabilities that surely, language development cannot be as important as I claim.

In response, I would ask you to consider a child who possesses no language; in fact, he is unable to communicate at all. The fictional child we are discussing here, (although there are many non – fictional children such as this), suddenly develops a temperature and is obviously distressed. Ultimately, the doctor is called. The doctor arrives, surveys the visible symptoms, concludes that the child is suffering from a generalised viral infection, prescribes medication and leaves. Two days later the child is hospitalised. Unknown to the doctor, he was suffering from appendicitis, or he could have been suffering a kidney infection; - it could have been one of a hundred things, which through no fault of his own, the doctor missed, because the child was not able to indicate where he was hurting. – Communication is *vitally important, - it can mean the difference between life and death.* Also, consider the importance of communication and language to two other issues, *motivation* and *understanding,* and the role these two play in learning and development.

A parent carrying out activities, which are designed to stimulate their child's development, is faced with two problems. Firstly, how do they motivate their child to want to do the things, which must be done? Well, we need to throw out the notion that children and their brain cells can be *programmed* in the way a computer can be programmed. – I have found no reliable evidence to support this notion. (There is a difference between 'programming' a brain and encouraging 'plasticity.' We can evidentially prove the plasticity of the brain, (it is called *long-term potentiation* and conversely, *long term depression),* whilst there is no evidence that a brain can be programmed with sensory – motor, or any other type of information). We need to abandon the view that a brain can be programmed and try to encourage the view that children are equal partners in their development, - that a programme of developmental activities is carried out *alongside them,* with their cooperation. Children cannot be 'programmed;' they are living, breathing, vibrant beings who possess their own minds and who deserve respect for that fact. If I should use the word 'programme' within this text, it does not in any way denote the type of 'programming' to which I allude here, I am simply referring to a schedule of developmental activities, which are carried out with the child's cooperation.

The motivation our children need to carry out developmental 'activities' can only come through communication and through language: - If language and communication are not present, we must attempt to develop them through playful interaction. It is unusual for a child without even the slightest communicative abilities to be able to resist play for long, if it is presented at the correct intensity and level of ability.

My second point is that of *understanding.* Again, this relates to motivation. The more a child understands, the more effectively he can be motivated. In turn, the most effective way to increase understanding is through language. This is an example of how learning leads development. If a child understands the need for him to carry out a task, which in turn will improve his developmental abilities, then the easier our task becomes.

The path of language development

Communicative competence at birth is basic; - baby has different types of cries to express different needs; a cry for hunger, a cry for pain, a cry for tiredness, etc. So even at this tender age, baby is producing meaning. What he also possesses is the initial tool, which will drive language development forward: - he is attracted to human faces! Why should this attraction to faces act as a catalyst for language development? Well, what do human faces do? They communicate by way of facial expression; - they also talk! Baby very quickly notices this and the adult adapts her interactive style to make communication more attractive to baby. She does this by speaking in a high pitched tone of voice; - by making communication jovial and an enjoyable, playful experience; - by attributing meaning to any sound baby will make and by shaping the talk of the adult and any noise baby may make into a turn – taking scenario. This instils into baby one of the primary rules of conversation right at the beginning: - These contrived turn taking sessions are called '***proto-conversations***.'

> *Proto-conversations: - These are the conversation-like interactions between an infant and a caregiver, in which the adult responds as if the intention of the baby's behaviour and sounds are understood. In this way, baby begins to appreciate the effects his actions have on other people and subsequently learns to adjust his activity according to the situation, context, and the actual intention of his interaction. Studies of proto-conversations between babies and caregivers have indicated their importance to the process of language development and socialisation.*

Is it any surprise that baby responds to all this encouragement and stimulation the way he does and that soon, the mere attraction for human faces has developed into a shared attention between adult and baby, which is called **primary intersubjectivity?**' Is it any surprise that baby soon becomes a willing and active participant in these enjoyable, playful, communicative episodes?

> *Primary intersubjectivity: The development of shared attention and understanding within a communicative episode, between baby and caregiver.*

This apparent drive towards co-operative interaction with others seems to begin at an early age and is seen in the *synchrony* of these early proto-conversations. The co-operative skills involved in the turn taking of proto-conversations lay the foundations for later language / conversational abilities. Indeed research has shown that not only do babies engage with their social partners with *mutual contingency* from shortly after birth, but also, from as young as six weeks they are attempting to interact in a coordinated, co-operative fashion. This is demonstrated in the study where the communications between mothers and their six to twelve week old infants were uncoupled. When mother and child interacted normally over a video setup, communication was responsive, but when the mother's video feed to baby was from a previous interaction, the babies acted in a more detached manner and there was no synchrony. (Trevarthen; 1995; Murray and Trevarthen, 1985).

Synchrony: - A situation whereby communication between two individuals is coordinated and in harmony.

Mutual contingency: - A situation where both partners within a communicative episode, adjust their level of response to coordinate with the previous communicative action of the other.

Due to the nature of their brain-injuries, many children are born without this inborn attraction to the human face. This places them at a severe disadvantage in their ability to develop the early cooperative abilities we have just described. Neither do they develop synchrony or contingency abilities, which would help towards later language development.

This situation is further compounded by the fact that many parents, because of dealing with the stress and tribulations of adapting to and caring for a child who has very serious problems, do not react to their babies in the way I have just described and so baby's opportunities to develop the early skills, which are necessary to language development are drastically reduced. This is not the fault of the parents, or other family members; it is caused by the stress, trauma and abnormality of the situation in which the family finds itself.

Difficulties may arise at any stage on the developmental pathway of language, causing an assortment of problems. The important thing to remember is that only through developing language and communication can we increase understanding, which in turn will increase motivation.

There are many other complications, caused by brain-injury, which affect language development. For instance, many children experience huge difficulties with muscular incoordination, so whilst they may understand much of what is being said to them (this is called ***linguistic competence***), they may not have the muscular control to produce understandable language, (this is called ***linguistic production***).

Hearing is also a major factor affecting the understanding and production of language. If the child suffers auditory undersensitivity, he may not hear certain ***phonemes*** (individual speech sounds) of the sound system of the English language; - consequently he will not be able to produce those sounds. Similarly, if he suffers from auditory oversensitivity, he may find hearing some speech sounds painful and be attempting to 'block them out.' Again, as a consequence the child will have difficulty reproducing these sounds.

If we consider the implications of brain-injury, then it becomes obvious that there are several reasons why a child with brain-injury may not develop the ability to share attention. For instance, many children have visual or other sensory difficulties, which would affect their ability to develop primary intersubjectivity.

By the age of two and a half months, children are beginning to make more noises than simply crying. The first one is what would be described as a ***cooing*** noise. There is also plenty of mutual smiling now between parents and baby and it is at this stage that proto-conversations begin to develop.

By three months of age, babies are crying far less. They begin to produce ***velar*** and ***fricative*** sounds and by four months, the early 'cooing' sounds are far more repetitive than before. They are also beginning to laugh and chuckle.

Velar sounds: - Velar sounds are made when the back of the tongue is pressed against the soft palate. They include the /c/ in cat, the /g/ in girl and the /ng/ in hang.

Fricative sounds: - Fricative sounds are created with almost a complete closure, but with just enough of an opening to create turbulence in the airflow. Examples of fricative sounds are /f/ as in fat, /v/ as in vat, /s/ as in sip, /z/ as in zip, and /S/ as in ship.

At five months of age, the repetitive 'cooing' noises begin to fade and the infant enters a period of ***vocal play***, where he experiments with a range of sounds. This is not yet so well organised as to be able to be called ***babbling***, but it is the child's first real attempt at using a semi-organised sound system. Soon after this, at approximately six months old the child begins to ***tone glide***. He also develops the ability to make ***nasal*** and ***plosive*** sounds.

> *Nasal Sounds: - /m/ /n/ and /ng/ are nasal sounds. The initial sounds in "mitt" and "knit", as well as the sound at the end of "bring", are called "nasals". That is because when we make these sounds, we push air out of our nose instead of our mouths.*

> *Plosive Sounds: - Plosive sounds are made by forming a complete obstruction to the flow of air through the mouth and nose. The first stage is that a closure occurs. Then the flow of air builds up and finally the closure is released, making an explosion of air that causes a sharp noise. The plosive sounds are /b/ /t/ /g/ /d/ & /k/*

At seven months of age, the child begins to ***babble***, although at this time the babbling will be limited to ***reduplicated*** sounds. This means the same sounds are repeated over and over again, such as '*Mamamamamama.*' This is soon followed at about eight months of age by ***variegated babbling***, where differing sounds are brought into the utterance, such as '*Badamaba.*' At nine months old, baby begins to introduce melody, rhythm and tone of voice into the babbling. At around this time, baby begins to develop ***secondary intersubjectivity***, which is most obvious with the emergence of ***triangulation***.

> *Secondary intersubjectivity: - The infant "shares" attention with a partner, with respect to some third (hence triangulation), external object or event. It has been claimed that the development of this ability shows that the infant is developing an understanding of the intentions or attentional interests of their communicative partner (Corkum & Moore, 1995; Tomasello, 1995).*

By the time baby reaches his first birthday, regular, predictable utterances begin to be made, which have clear meaning and are in context. This period is notable for the child's use of ***scribble talk*** or ***jargon***. This is where baby makes many varied noises, which are very 'word like.' This is a precursor to baby uttering his first words, which usually occurs at around fourteen months of age. After this, new words are added to the vocabulary regularly and language production improves as understanding of the grammatical structure of the language improves.

It is vital that parents encourage the development of as wide a vocabulary as possible in their children, - that they talk to them as much as possible and encourage their children to talk. There is evidence that language use and vocabulary power in young children, leads to later scholastic advancement. - In this sense, learning occurs through talking. (Corson, 1988).

It is vital that our children are taught to use language in this way as it then becomes a sociocultural tool, which can help develop a social mode of thinking. – By this, I mean that the child is able to utilise language, in his communications with others, such as parents, in what are social learning situations, in order to further his own development. This is because learning is largely an 'external' social activity, which then becomes internalised to become part of the child's development.

Problems of language and communication

There are many forms of language and communication difficulties, some of which can be misleading, leading the professional to conclude that the child's problems are not merely to do with language, but that they have a totally different problem, such as autism.

I was called into a small local primary school by the head teacher, who was convinced that a young girl in her class was suffering from autism. The girl in question was reticent, socially isolated, keeping herself well away from others and her language and communication skills were severely retarded. It took three sessions of observation to draw the conclusion that this was definitely not autism! It turned out that the young lady in question was suffering an SLI (Specific language impairment). Her poor communication abilities were having such a dramatic effect upon her morale that she had simply retreated into herself and stopped trying to communicate. I suppose if you fail every time you attempt to communicate with spoken language, in the end you simply stop trying for fear of further failure. It took two years of patient painstaking teaching, taking the young lady back through the earlier stages of language development to achieve a situation, where today she is a bright, happy, socially active young lady, whose social and academic prospects are now normal!

There are many other patterns of brain-injury, which can cause a myriad of difficulties with the development of language and communication. Children may experience these difficulties as 'stand alone' problems, or they may experience them as just a part of a more complex set of issues, such as autism. Such difficulties include, speech production difficulties like ***Broca's aphasia***, where the child produces sluggish, halting speech, or ***Anomia,*** which is characterised by word – finding difficulties. Then there is ***verbal apraxia***, where the child displays difficulties positioning speech sounds in the proper order, so as to create words.

A difficulty of speech production common to children with brain-injury is ***dysarthria***: - a disorder in which the muscles, which are utilised during speech are weak and badly coordinated. Therefore, speech is slower, imprecise, incoordinated and slurred.

Some of our children also suffer speech comprehension difficulties, such as ***Wernicke's aphasia***. This is a set of difficulties where the child displays deficient speech comprehension and produces incoherent speech. It is caused by injury to the posterior part of the frontal lobe. Meanwhile, injury to part of the temporal lobe can cause a phenomenon known as ***pure word deafness.*** This condition reveals itself as the incapacity to understand the meaning of spoken words.

Yet another type of aphasia is ***transcortical aphasia***, of which there are three types; - mixed, sensory and motor. The child who experiences these difficulties has a tendency to repeat words to the point of sounding like an echo.

Children who suffer from ***conduction aphasia*** are not capable of repeating words spoken to them. Spoken language is comparatively steady, although the child might regularly correct himself and some words may be omitted or reiterated. Even though they are able to understand spoken language, it can be challenging for the child who suffers from conduction aphasia to locate the precise word to express their thoughts.

Another interesting type of aphasia is ***paraphasia,*** which is characterised by the construction of inadvertent syllables, words, or phrases throughout a child's spoken discourse. Children with types of aphasia where spoken language is smooth display many more paraphasias than do those whose aphasia produces broken, halting speech.

Finally for the aphasias, from which our children may suffer, there is **_anomic aphasia_**. The child suffering from anomic aphasia has immense difficulty in recalling the names of familiar objects. For example, presented with a banana, he will say **_'oh yes, it's a , a, a, - yellow thing, you peel it and eat it.'_** They are not able directly to label it a banana! We have all experienced this phenomenon, where the name of something is just maddeningly out of our grasp. These children suffer the situation constantly.

Mobility development and the reflexes

Problems affecting mobility development

Many children who suffer from autism do not have significant mobility difficulties; - they can make reasonable progress under their own steam. There are a minority however, where the brain-injury is more complex, who do experience problems with mobility development. It is to parents of those children that this section is addressed.

Most people, even those who have little experience of children, are familiar with the major stages of mobility development. Most of us know that first, baby holds his head up, then later he rolls, sits up, crawls (first on his tummy, later on all fours), then he stands and finally walks. Many of us have seen this process; - those of us who have not, probably know the sequence. It is in the area of mobility that the difficulties caused by brain-injury often become most obvious and most open to public scrutiny. We have all seen children in wheelchairs, children in standing frames, callipers, crutches etc. Lack of development in this area can be one of the first outward signs that many parents become aware of, regarding their children's difficulties.

In the type of brain dysfunction, which causes autism, there could be many areas of the brain, which may have been injured, (a *diffuse* injury), which depending upon the complexity of the injury, can affect the child's ability to move.

There are definable parts of the brain, which when injured produce impairments of movement and hamper mobility development. For example, damage to parts of the brain called the *motor cortex,* or the *pyramidal tract* produce muscular stiffness and mobility problems. Injury to the *corpus callosum, frontal lobe,* or *parietal lobe,* cause a phenomenon known as *apraxia,* which literally translated means 'without action.' This is a phenomenon where the child simply cannot initiate or imitate a movement, or produce a movement on request. There are four major types of apraxia, which may affect a child with autism. First as I have just described there is *limb apraxia.* The second affects the mobility of the speech muscles, which may cause your child to suffer from *oral apraxia.* This is where the child has difficulty in utilising the muscles, which make it possible to speak. Thirdly, the child who experiences very mild brain-injuries may experience ***apraxic agraphia,*** which is a type of writing difficulty. Finally, and again this would only be noticed in a child with very mild brain-injuries, there is ***constructional apraxia,*** which displays itself by the child experiencing difficulties in drawing or building models. (Carlson, 2007).

The path of mobility development and the reflexes

At birth, children are born with many postural reflexes, but as a child grows and develops, these reflexes become overlaid by normal bodily movement patterns. In many brain–injured children however, the reflexes are either displayed way after the time when they should no longer be apparent, or they are not present at all.

One of the first signs that mobility development is destined to be problematic is the child being late in achieving head control (some children who have suffered profound brain-injuries never even achieve this). Head control appears early in mobility development because there is a developmental process at work which is known as ***cephalocaudal*** development.

Cephalocaudal Development: - A pattern of physical development, which begins at the head and gradually moves down the body, through the neck, shoulders, trunk etc.

Another aspect of physical development is that it is also

proximodistal,

Proximodistal development: - A pattern of physical development, which begins at the central parts of the body and gradually moves outwards until baby can control hands and feet.

Bearing in mind these two developmental processes, it is not a difficult task to judge whether physical development is progressing as it should. Mobility development is also closely linked to tactile development. If a child cannot feel his own limbs and cannot feel his environment then how can he possibly move within it? In fact, it would be dangerous to try to do so.

At birth, a child is born with very little mobility. He produces **whole body movements**, which are essentially squirming, wriggling type movements, and flails his limbs in what appears to be a random manner. He possesses several reflexes, such as the **eye-blink** reflex (to elicit this, quickly move your hand towards his face, stopping just a couple of inches short), and the **suckling reflex**. (The latter is necessary for survival). You will also notice that if you hold his hand, he will pull the limb away from you; - This is not voluntary, it is called the **withdrawal reflex**. – This should disappear after he is about two weeks old.

You will also have noticed that if you hold baby close to your chest, he will 'root.' This is where he will nuzzle into your chest and shake his head from side to side. This is called the **rooting reflex** and is an attempt to find the nipple and as a result, food! This should disappear when he is about one month of age. At this young age, if you hold baby in an upright position so that his feet are in contact with the floor, you will be amazed that he will try apparently to walk! – This is called the **stepping reflex**. This phenomenon should disappear after approximately two months. Most, but not all babies exhibit this reflex. An important point to remember here is that baby cannot support his own weight at this stage of development and should never be expected to bear weight at this age, so **do not** attempt to place him in a situation where he is expected to stand upright and take his own weight! There is evidence that if the stepping reflex is stimulated, with the infant given stepping practice, (whilst an adult supports **ALL** of baby's weight), the outcome for the child is that he will walk earlier than he normally would. (Zelazo & Kolb, 1972).

Baby remains relatively helpless in physical terms until about four months of age, when head control begins to improve. At this point, the **tonic neck reflexes**, should have disappeared. – To highlight these reflexes, if you were to lay your young baby on the floor (on his back), you would notice that his head falls to one side. You should also notice that he holds his head and arms in specific positions (he looks rather like an archer holding a bow) and that when you gently move his head so that he is looking towards the other side of the room, his arms and legs adopt exactly the reversed positions. These postural reflexes disappear at about four months of age in the uninjured child.

At about five months of age, you may notice that the child starts to try to roll. The initial 'side to back' rolling can begin as early as two months of age, but proper rolling begins at about five months. Rolling is often the first real independent movement by the child and initiates the vestibular stimulation the brain requires in order to help to develop the mechanisms, which will control balance and coordination. Many children who suffer autism, have poorly developed vestibular systems and this is a further constraint on the development of mobility.

At about six months of age, the *moro reflex,* (otherwise known as the startle reflex) should disappear. The moro reflex can be elicited by lying the baby on his back and creating a loud, unexpected noise; - or by making baby think he is falling. He will stiffen his back, throw his arms out wide, with his fingers splayed and his arms will shake.

In many children who have brain-injuries, all of the previously mentioned reflexes may stay with the child well after the time at which they should have disappeared. Of course, on the opposite side of the equation, many or all of the reflexes we have discussed may not be present at all in the child with brain-injury, or may not show themselves at the appropriate time. This again is due to the effects of the brain-injury.

It is at about six months of age that the child normally begins to sit up. This is an important milestone, which leaves baby's hands free to explore the immediate environment. It is obviously no co-incidence that the *grasp reflex,* which baby possessed at birth, disappears in the weeks prior to this development. To demonstrate the grasp reflex, place your finger in the palm of baby's hand. He should wrap his fingers tightly around your finger and he will be unable to let go. It has been argued that this is an evolutionary throwback to the times when as apes, we needed to cling to our mothers as we journeyed through the trees.

The *body righting reflexes,* which have been operating for some time, are now obvious and are utilised to keep the body and head correctly oriented in space.

The five body righting reflexes and their functions are as follows:

1) The Labyrinthine-righting reflex: This maintains the head's orientation in space.

2) The body-righting reflex: This keeps the head orientated to the body.

3) The body - righting reflexes from the body surface receptors: These orientate the body in space.

4) The neck - righting reflexes: These keep the body orientated to the head.

5) The optic- righting reflex: This keeps the head in proper orientation in space.

At seven months of age, the child begins to develop the capacity for

independent forward movement. This is haphazard and laboured at first but quickly becomes more efficient and purposeful. At first, this is achieved through the process of ***commando crawling,*** which is where the child lies flat on his belly and pushes himself forward much as a true army commando would crawl on his belly. Soon, approximately one month later, baby switches to ***quadruped crawling,*** this is where baby crawls on hands and knees, with his torso off the ground. It is at this point that visually, depth perception develops and it is important that it does so, as baby is now moving at comparatively greater speeds and needs to be able accurately to judge the distance to and location of objects in the environment.

By ten months of age, baby has developed sufficient vestibular control to allow him to pull himself to a standing position whilst holding onto a piece of furniture, though he is not yet ready to stand alone. It is at this point that babies roam around the house using the furniture as a constant support; - it is what our cousins in the US term ***cruising***. It is only a few short weeks however, by the age of eleven months usually, that baby is able to stand-alone fully. After this, it is merely a short step to walking, which is usually achieved around the first birthday. It is now that the body righting reflexes and the ***babinski reflex*** disappear.

> *The Babinski reflex occurs when the big toe flexes toward the top of the foot and the other toes fan out after the sole of the foot has been firmly stroked. This is normal in younger children, but abnormal in older children. A continued Babinski response after twelve months can indicate damage to the nerve pathways, which connect the spinal cord to the brain.*

It is now baby is walking that you will notice another evolutionary phenomenon. Look at his arms and how he is holding them up, just as you have seen apes on television holding their arms when they walk upright. As he becomes a more skilled walker over the next few weeks and months, you will notice that the arms will slowly, but surely drop down, so that eventually, he is walking as we walk.

As his abilities improve, the young child is able to add to his repertoire of skills and by eighteen months, he is able to walk upstairs largely without assistance. By two years of age, he is able to jump up and down! By three years of age, he is able to walk and run as fluently as an adult and by age five, the early signs of

hemispheric dominance, which began to be displayed when he was two to three years old are maturing and he now clearly shows which side of his body is dominant; - whether he is right or left handed and footed and is performing skilled activities with those limbs.

This is the developmental pathway, which we all, as human beings have traversed. Brain-injury can prevent our children from making the journey down this pathway. If we are to succeed with brain-injured children then we must find ways to facilitate their passage.

Fine motor development

Children with autism may present with many assorted fine motor difficulties. Early fine motor development may be in harmony with usual developmental norms. It is possible that problems may become apparent, with such activities as the use of cutlery, or the ability to manipulate objects in addition to difficulties in drawing and writing. Let's look at how fine motor development progresses in an uninjured child, so that we may know exactly what we are aiming for.

The path of fine motor development

Newborn babies are, in many quarters considered helpless, relatively inactive beings who need time to mature and grow in order to interact effectively with their environments; - That is the view sponsored by the students of Jean Piaget and is still an influential view within Western societies. In these pages, where I have highlighted normal developmental pathways, I have indicated that this Piagetan philosophy simply is not applicable to the ways in which children think, learn and develop. Newborns are complex, sophisticated beings and this remains the case as far as fine motor development is concerned.

At birth, baby possesses a grasp reflex; - we have already established the evolutionary importance of this, but were you aware that newborn babies can and do actually reach for objects? This phenomenon is called ***pre-reaching***. Were you to lay a baby on the floor, on its back and hang a small object, centrally above baby's body, but within reach, you would notice some definite attempts to reach for it. They may be clumsy, disjointed attempts, but they are definitely there and have been noted by many researchers. (Berk, 1997).

Within two months, the pre-reaching of the newborn disappears and voluntary reaching begins. These movements are far more delineated and purposeful, although still quite clumsy. Baby has now lost his grasp reflex, which has been replaced by what is called an ***ulnar*** grasp, which is a clumsy motion, the child's fingers closing against palm of the hand. There is no finger and thumb opposition at this stage.

At four to six months of age, baby is beginning to manipulate objects and is starting to transfer them from hand to hand. Also at around six months, he will not bother trying to reach for objects, which are clearly out of reach, indicating that there is a rudimentary appreciation of depth and the third dimension.

By nine months of age, fine motor abilities are becoming more sophisticated and baby can even redirect his reaching to reach for a moving object, which has changed direction. By ten months, he has finally achieved finger and thumb opposition in what is known as a ***pincer grasp.***

At twelve months of age, baby is beginning to point at objects accurately and is now able to pick up objects using what is now his sophisticated pincer grasp, which by fifteen months of age has become a very precise grip used with either hand. Baby is now able to manipulate toy building blocks quite dextrously and may try to build a tower with two or three blocks. At this age, the youngster is able to grasp a crayon with either hand, but interestingly reverts to an ulnar grasp when doing so. He will also attempt to hold a spoon and feed himself.

By eighteen months of age his tower building skills, using cubes has progressed to successfully utilising three or four blocks and he is now more successful at feeding himself with a spoon. He can also now hold a cup in both hands and give himself a drink without spilling it.

At two years of age, he possesses good manipulative skills and is able to pick up tiny objects. You will notice that now, when he holds a crayon, he holds it well down the shaft, correctly gripping it with his thumb and first two fingers. At this age, you may also detect that he occasionally demonstrates a preference for using left or right hand, - an early indication of a developing hemispheric dominance.

By three years of age, he is able to use two hands in a cooperative manner in order to complete a task and his tower building skills have now progressed to using approximately a dozen blocks. He is now showing clear signs of preferring one hand to the other and is able to draw a person, which will possess a head and one or two other features.

By four years of age, he is able to build complex structures using blocks; He now holds a pen as an adult would hold it, in his preferred hand and is able to make quite complex drawings with it. – This improved control of the writing implement is further augmented by the age of five, when he shows good control in both his writing and drawing. His drawing of a person now has all body parts incorporated and he is able to colour pictures neatly. At this age, the child is now demonstrating clear and unequivocal superiority of one hand over the other, indicating that the brain is well down the road to developing a dominant hemisphere. *(Lateralisation)*

Social development and the sense of self

This is an area, which characterises the entire concept of autism. To many parents the lack of willingness on the part of their autistic child to share in normal social action is of paramount concern. One parent described her child as having **social amnesia.**

The social impairments, which typify autism are exact, that is, the child's social conduct is not atypical universally. It is incorrect to declare that children, who are autistic, have a deficiency in their level of curiosity in other people. What they are deficient in is the proficiency for conveying or exploiting that interest. Uninjured babies are focused on faces and voices, whereas autistic children do not seem to be! They do not turn automatically to the sound of a voice, or fix their eyes on a parent's face, and may actively avoid meeting their vision. In many cases, this is due to sensory impairments, which can block the development of these social skills.

We will take a look at how the development of social skills takes place and look for answers as to why our children do not develop socially.

The ability of the infant to co-operate in interacting with others is instrumental in developing a sense of self and of others as being separate from self and it is an ability, which is often lacking in children who suffer autism.

The drive toward co-operation with others seems to begin at an early age and as we have discussed, it is seen in the **synchrony** (cooperation) of early proto-conversations. The co-operative skills involved in the turn taking of proto-conversations are learned through early interaction with parents and lay the foundations for later language / conversational abilities. In itself, this turn taking routine marks some rudimentary sense of separateness from others. Indeed, research has shown that not only do babies engage with their social partners with mutual contingency from birth, but also, from as young as six weeks they are attempting to interact in a coordinated, co-operative fashion. (Trevarthen; 1995; Murray and Trevarthen, 1985).

Despite this early co-operative behaviour, some researchers still claim infants have no sense of self in the early months. The position taken by these researchers comes from evidence gained from the **'mirror–nose test'**, where it was only by the ages of fifteen to twenty - four months that infants in fact did reach for their own, 'rouge painted' nose, instead of their reflection, indicating their realisation that the reflection was themselves. (Lewis & Brooks – Gun, 1979).

However, there are many displays of co-operation by young infants, which indicate that the sense of self and other people may be more sophisticated than these researchers claim. Infants as young as six months have been known to produce 'checking behaviour,' co-ordinating their own view of a situation with their carer's view of the same situation. This surely indicates a sense of self as being 'apart' from others. (in this instance, the carer). They will also engage in 'play fighting' and demonstrate fear of strangers. This indicates an awareness of being 'apart' from the person being play – fought and latterly, 'apart' and different to the stranger of whom they are wary. Unfortunately, the effects of brain-injury can preclude the opportunity for this development of the sense of self and 'other.'

A year old infant can adjust to take in the perspective of another person, in order to co-operatively share an experience; 'perspective taking' being a central tenet of many accounts of development. – Its necessity derives from social interaction, (which is itself a co-operative endeavour), which further promotes it. The interesting question is whether this seeming drive towards co-operative social interaction is causal in promoting the development of a sense of self and sense of others? The evidence suggests that it is indeed the co-operation within these early interactions, which triggers the development of a sense of self and 'other' as being separate entities. Some children with brain-injuries tend not to be involved in such early co-operative interactions for two reasons.

(1). The nature of their developmental problems prevents the child from interacting appropriately. – This is certainly the case as far as children with autism are concerned.

(2). Parents of children who have brain-injuries tend not to interact with their children as parents interact with uninjured children. (This is not the fault of the parents, but a consequence of the immense stress, which naturally comes with adjusting to the child's difficulties).

Consequently, we have a situation whereby the very means of development (normal interaction with parents) is denied to this group of children.

Research has shown that the sense of self / sense of other undergoes further development in the second year and that it is the child's understanding of the emotional states of others, which precedes understanding in other areas, such as understanding the mental states of others. (Stern, 1985).

It is at this time the child begins to develop ***internal working models,*** of both themselves and others. The types of models developed by the child are again linked to the word 'co-operation.' Secure (co-operative) attachment experiences and relationships tend to produce the development of positive internal working models whilst insecure (uncooperative) attachment experiences cause more negative models to be developed. Here again, the situation appears to be that 'co-operative' social experiences are moulding the sense of self and the sense of 'other.'

Unfortunately, due to the difficulties which surround the child's problems, children who suffer autism cannot develop their co-operative skills and therefore do not form the secure attachments of which I speak.

> *The internal working model: - This is a mental representation of either oneself or other people. It comprises physical appearance and personality attributes.*

This view is supported in many studies where it has been demonstrated that infants who experience secure (co-operative) attachments develop more self-awareness, more featural knowledge of themselves and others, higher self esteem and better self-concept, than children who experience insecure attachments or who are maltreated. (Does not autism maltreat a child?) (Pipp, Easterbrooks and Brown, 1993).

The development of language has a dramatic effect upon the child's ability to co-operate, both through discussion, which is a co-operative activity in its own right and through co-operative and pretend play. It is at this point that children are considered co-operative enough to become what Rogoff (1990) describes as ***apprentices in thinking.***

> *Apprenticeship in thinking: - This is a point in time where children develop their thinking abilities as a consequence of active participation in activities under guidance of their caregivers.*

Indeed, at this point children gain a fresh sense of self-awareness, brought forth by the advances the child makes, both in linguistic competence and production. This new sense of self is most evident with the child's use of self descriptive terms such as 'I,' 'me' and the use of his own name. Again we see that a 'co-operative situation, facilitated this time by language, enables the child to make advances in self-awareness. (Schaffer, 1996).

The importance of play

One of the first signs that a toddler or preschooler has autism is their atypical play. Even the brightest youngsters with autism display highly unusual patterns of play. Classically, children with autism over-focus their attention on visual aspects of specific toys, or noises, which their toys make. Many researchers see this as a lack of imagination in autistic individuals and it is true to say that children with autism do lack imagination and spontaneity within their behaviour, preferring to stick rigidly to routines with which they feel comfortable and safe. However, what I propose, is that many times, these problems are created as a result of the sensory distortions, which they suffer. We have already discussed such sensory distortions, which produce these behaviours in many autistic children.

When children play together, as they do in a more and more co-operative fashion after the age of three years, they are demonstrating a range of skills. During play, they are engaging in a highly co-operative way and demonstrating a sophisticated level of intersubjectivity. One of the most important venues for children to interact in this way is at preschool, which provides excellent opportunities for them to utilise and sharpen these skills by providing the opportunity for them to engage in co-operative and pretend play with their peers. However, many children who suffer brain-injuries, particularly the type which express themselves as autism, even if they have by some miracle managed to develop this far in social terms, are not afforded the appropriate opportunity, or due to the severity of their problems are unable to engage in pre-school activity.

One of the features of play in children in their third year is the way in which their pretend play, which until now has been based within a role of compliancy, becomes more social and shared with other children. There is evidence that the nature of pretence in play is to allow the child to gain an understanding of his own emotional life. It has been demonstrated that children's representations during pretend play originate from significant emotional experiences in their lives. (Dunn, 1991).

The very act of pretence requires a sophisticated knowledge of both self and other people, because the child must demonstrate the capability of taking on the role of a pretend person who may have emotions and thoughts which may not coincide with his or her own. This clearly shows the child's growing ability to understand the mental states of others and is supported by other evidence, such as the fact that during the second half of the third year, children do begin to speak about their own and other people's mental states. (Dunn, 1991).

In addition, the level of intersubjectivity and **metacommunication** witnessed in pretend play scenarios requires colossal co-operation between playmates. The co-operative play and pretend play fostered by the 'preschool' situation is of great benefit to children as they gain the opportunity to sharpen the skills, which will enable them to achieve a better understanding of themselves and others. To highlight this point it is interesting to consider the fortunes of children who have difficulty with co-operative and pretend play.

Children have a need to be able to understand the perspective of others in order accurately to predict their responses, in addition to guiding their own actions; - Many researchers call this ability **theory of mind** and it an ability, which is notably lacking in many children who suffer autistic type brain-injury.

Metacommunication: - The negotiation of the direction of pretend play by two or more play partners. Also the negotiation concerning symbolic objects and their use, i.e., using a saucepan to represent a helmet.

Theory of mind: - An understanding of the perspective and intentions of other people, used to predict what another person might do or say next.

There is evidence to support the notion that children who are part of a larger family develop these abilities earlier than children from smaller families. It seems to be co-operative interactions with siblings, which facilitates the child's attention being focussed on the state of mind of his / her siblings. This enables the child to develop insight into their mental state. (Schaffer, 1996).

There is also evidence that the development of 'theory of mind' is dependent upon children engaging in pretend play. The question, which needs to be addressed is; - is the fact that so many children who suffer from brain–injuries (such as those displayed in autism), do not engage in pretend play, the reason why they show a deficiency in the development of theory of mind? Obviously, there will inevitably be a large proportion of children whose severe physical and intellectual limitations automatically exclude them from participation in pretend play and this must be detrimental. - It is an interesting question.

It has been suggested that the ability of children to understand the minds of others is influenced by interaction within the family. In a study of families in Cambridge, correlations were found between the frequency and extent of family interaction and the later ability of children to understand the minds of others. As I have already pointed out, children who suffer brain-injuries are often not afforded the opportunity to interact in this way, - or they are prevented from doing so by the severity of their problems. (Dunn, Brown & Beardsall, 1990).

Around the third year, further evidence for children's growing ability to understand the minds of others can be seen within family interactions. Consider the evidence from one family where a child in her third year had an older, only sister who possessed imaginary friends. In order to upset her sister, the child would transform her own identity and claim the identity of one of the imaginary friends! This shows an understanding not only of herself as a separate entity, but also of the attributes of the imaginary friend. It also demonstrates a great deal of determination and imagination to infuriate her older sister! (Dunn, 1991).

It follows that further co-operative interaction, facilitated by siblings and peers in family and preschool environment, particularly the opportunity afforded for the child to participate in co-operative and pretend play, further augments the child's growing knowledge of self and other people.

Having highlighted the major problems, which may affect your child in terms of his development, it is now appropriate to move on to what we can do in order to combat those developmental problems. Consequently, we move on to 'treatment.'

Treatment

Many in the health professions, still adhere to the old maxims that 'once a brain is injured, there is little that can be done to change the consequences, which ensue' and that 'the brain and its functioning are fixed and unchangeable.' There is a great deal of evidence which contradicts this view and supports the idea of a malleable, plastic brain, whose developing structure and functional capability is dependent upon the environment to which it is exposed. There is also evidence that new neural pathways can be created in an injured brain, which can assist in recovery of function after brain-injury. Let us explore these issues in a little more depth.

The simple fact is, the human brain possesses a high degree of ***plasticity***. The younger the person is, the greater plasticity the brain may exhibit, but even in older people the brain has demonstrated surprising levels of plasticity. The brain has a great deal of physiological reserve and in some circumstances can tolerate loss of neuronal function. (Slater, 1995).

> *Plasticity: - The ability of the brain to adapt itself to altered circumstances, by changing it's structure and function, whether this is in response to injury, the need to learn a new skill, or to a changed environment. – To find new ways of learning within that changed environment.*

Evidence for this is demonstrated in the work of Elbert (1998) in which he demonstrated that the brains of people who were trained in the new skill of learning to play the violin, assigned new neural elements which were dedicated to controlling the new ability. The phenomenon has also been demonstrated in people who were learning to read Braille for the first time. (Sterr, 1998).

Every day, in all adult individuals, some brain cells die. This is a process known as ***apoptosis.*** So why do we not all slowly lose our functional abilities? Well, the answer lies within the concept of plasticity. As our brain cells die, the plastic qualities of the brain determine that the physiological reserve of the brain is diverted to the task of ensuring those functional capabilities remain intact. The plasticity of the brain ensures that as fully functioning human beings, we continue as before. To illustrate this, consider the fact that neuroscience has discovered that during a person's lifetime, as brain cells die, there is growth in ***dendrites*** in areas of the brain, which can only be explained by healthy brain cells taking over the functions of their dead neighbours.

There is also an increase in the size and number of certain brain cells, which are capable of releasing chemicals, which promote the growth of new neurons. – This is known as ***neurogenesis***. (Slater, 1995. p. 82).

> *Dendrites. –Short, branching fibres, which extend from the cell body or soma of the brain cell. These fibres increase the surface area available to the cell to receive incoming information.*

However, consider children who suffer from autism and many other expressions of brain-injury. Why is it, do you think that the brain's quality of plasticity does not ensure that the functional abilities which were lost as a consequence of the brain-injury are quickly recovered? Well, instead of just a few brain cells being lost daily to natural causes, as is the case with you and I, during brain-injury millions or even hundreds of millions of cells are lost in one brief but terrible insult to the brain. Whereas a few cells are relatively easy to replace, (indeed it seems the brain must have cells on 'standby' ready to take over the task of cells, which die), it is much more problematic for the brain to replace the functional ability, which is lost when hundreds of millions of cells and their associated connections are wiped out simultaneously. (Indeed, entire functional areas of the brain may be lost in such an injury, if the injury is localised in one area).

Even in the drastic circumstances of brain-injury, the plastic qualities of the brain do ensure that some functional ability is usually recovered, (except in the most severe cases). However the important question is; - How do we maximise the plastic qualities of the brain and consequently promote the maximum recovery of function, (or development of function), in children who have suffered this type of catastrophic brain-injury?

We know that one thing the brain does very effectively and naturally is to ***learn***. We also know, from the work of Vygotsky that learning can lead development. Although the learning capacities of an injured brain may be impaired, if we can harness the inherent plasticity of the young brain, by providing it with the optimum neurological environment in which to learn, then we can have a chance of producing dramatic functional change even in severe brain-injury.

The 17th century philosopher, John Locke proposed that the mind of a child is a 'tabula rasa,' (a blank slate), and consequently, given the correct environment and guidance, the child's potential is virtually limitless. Well, although this is not quite the case, there is some evidence to support the claim that the child has a huge amount of potential. It seems that although we must not underestimate the role, which genes play in the development of the brain and of the child, it is the environment to which a child is exposed which is still the most powerful determinant for the child's developmental prospects. We know from studies where children have been discovered in impoverished circumstances, that consequently, they have suffered severe developmental problems. We also know that the provision of a well designed, highly structured environment can positively influence development. This is as true in children who do not have brain-injuries as it is in children who do!

There are many facets to treatment, but the primary problems, which need to be resolved, are the problems of sensory perception. If a child is unable to perceive his world correctly, then he will not be able to respond correctly. ***It is that simple!*** I have indicated that many output abilities, such as language, mobility and fine motor development are subservient to sensory functions. For instance, it would be difficult and dangerous to move, without the sensory feedback provided by vision and touch. Spoken language does not develop properly in the absence of hearing. Therefore, it is clear that sensory development precedes the development of output functions.

The question is, - what can we do to help sensory abilities develop? Well again, this is a matter of learning! Not conscious learning as a child might learn its multiplication tables, but a level of subconscious learning, which re-educates the malfunctioning brain. Every child I see as a client undergoes a thorough evaluation of his sensory capabilities. From this evaluation and from taking account of the testimony of his family and other professionals involved in his treatment, the child's level of sensory perception is established. - Is he over-sensitive? Is he under-sensitive? Does he have an 'inward tuning' problem? Once his sensory status has been determined in the areas of visual, auditory and tactile development, using Snowdrop's developmental profile, the next step is to devise a set of activities for the family to carry out at home, with the child. If the child has problems of sensory distortion, then activities are aimed at providing the correct neurological environment to enable his sensory system to normalise. This consists of creating an ***adapted environment,*** which can be a room within the home, the child's bedroom for instance, which can serve as a setting within which the correct sensory stimulation can be provided.

If the child has a 'straight' developmental difficulty in sensory terms, then activities are designed to stimulate the child to achieve the next higher level of sensory development. To highlight this we can again review our categories of sensory problems to highlight how to treat them.

Treating sensory distortions

Wide spectrum tuning

The child who suffers wide spectrum tuning difficulties experiences many sensory phenomena within the environment simultaneously. He may experience wide spectrum tuning in one modality, i.e., vision, but not others such as hearing and touch. Alternatively he may experience this problem in all sensory modalities. It just depends on what particular configuration the brain injury takes. – In the case of vision for example, his system seems incapable of selecting and attending to one visual stimulus, attempting to attend to many stimuli simultaneously. The result is most often sensory overload and a child who retreats into himself as a defensive strategy. So what can be done to help him?

Wide spectrum tuning: - Vision

For this child the answer lies in providing an environment, which is free of visual distractions. We must teach this child to focus his attention on a single stimulus to the exclusion of all others. Several times each day, this child needs to be placed in an environment, which has been specifically created to be free of visual paraphernalia; - no mobiles, no mirrors, nothing that shines or reflects light. Next, we must make sure any lighting within the room is subdued. We then provide him with short bursts of stimulation, which are designed to necessitate him focussing his attention on a stimulus of our choosing. (This stimulus can be a fluorescent toy, a coloured light, or merely the playing of a game like 'peep – oh,' which encourages eye contact).

Hopefully, as the situation improves, we can begin to introduce normal background stimuli into the sessions. This needs to be at a very low level initially, not so intense as to interfere with our encouraging him to focus his attention on the stimulus of our choosing. As his capacity to focus his attention on a single stimulus improves, the background stimulation can slowly be brought up to normal levels until he is able to focus his visual attention on a single stimulus in the setting of a visually normal environment.

Wide spectrum tuning: - Hearing

The child, who has problems with auditory 'wide spectrum tuning,' has a sensory attentional system, which is attempting to attend simultaneously to many sounds in his auditory environment. This is the child who lives in a state of chaos; to him the world is an overwhelming, confusing and frightening place. As with the child who suffers from visual wide spectrum tuning, this child needs the provision of an 'adapted environment' where we can attempt to re-tune his sensory system. This involves placing him in a quiet room where all extraneous background noise can be eliminated. He cannot filter out background sounds, as we are able to, so we must temporarily do this for him by providing him with a setting where background noise is simply not a problem. The child needs to visit this environment several times daily.

Whilst he is within this 'auditory cocoon,' we can begin to introduce one single auditory stimulus at a time, to attempt to encourage him to focus his auditory attention. As with vision, these sessions do not have to be terribly prolonged. Indeed, as soon as he shows the slightest disinterest in the situation, the session should be stopped. Remember, children do not progress if they are not motivated to cooperate. It is my experience however that your child should quickly react positively to this environment and show considerable relief at being cocooned within it. If he is relaxed and enjoying the adapted environment, it is good to leave him in it for a while, even if you are not working with him. This will help to calm his anxieties and allow him to enjoy the experience of not living in a world of chaos.

As your child begins to adapt to the controlled environment, it is possible slowly and very gradually to increase the level of stimulation within it. This must be done at a pace with which he is comfortable; - he must never be allowed to experience discomfort in this setting; - it is his safe haven.

Wide spectrum tuning: - Tactility

Children who suffer from tactile, 'wide spectrum tuning' seem to have immense difficulty in focussing on one particular sensation at any one moment, being simultaneously aware of many points of contact on their bodies. This is possibly the trickiest of all sensory problems to deal with, as the principles of treatment suggest we should create a 'dampened' tactile environment. However, short of actually levitating someone in mid-air so that they are not in contact with any surface or item of clothing, this is impossible to achieve.

So, we can only do what it is possible to do and consequently we should try to achieve as low intensity, 'general tactile environment' as possible. We should also simultaneously increase the tactile stimulus in a specific area, to enable him to focus upon a particular sensation in a particular place on the body. The same area of skin should not be used repeatedly however, lest it become sore and uncomfortable. Again, these exercises should be carried out as many times as the child will allow during the day, in the room, which we have created as an adapted environment for our child. As the child's sensory system re-tunes, we should be able gradually to normalise the intensity of the tactile stimulation we are providing.

Narrow spectrum tuning

Narrow spectrum tuning: - Vision

This child can seem obsessive about certain aspects of his visual environment; - certain toys, moving objects, etc. What seems to be an obsession is more often than not, an over-focussing of visual attention. Once his attention is gained, it is an 'all or nothing' affair. It seems he simply cannot easily shift his visual attention to other aspects of the environment. The strategy to combat this is again to create an artificial environment in a room in the home, where there are a few features, or points of stimulation, which stand out equally. If over-focussing occurs with reference to one object, that object should be withdrawn. His helper in this situation should also present him with situations where she /

he has two interesting stimuli, which she can 'switch' from time to time, to allow him to become used to 'switching' his attention.

Narrow spectrum tuning: - Hearing

This child over-focuses his auditory attention on specific noises within his environment and has difficulty 'shifting' that attention to other salient features. Again, the first phase of treatment involves the creation of our adapted environment in which the child can be placed, whilst we provide appropriate stimulation designed to re-tune his sensory system. The child is then to be placed within that environment several times daily. Within these 're-tuning' sessions, we must first provide an auditory stimulus that the child will find interesting and attractive; - this will give him a point of focus for his auditory attention. We must then introduce several competing stimuli in order to encourage the child to begin 'switching' his attention. The intensity of these competing stimuli must be steadily increased until they cannot be ignored. We must replicate this procedure, over and over again until the child can accept and acknowledge several features of his environment.

Narrow spectrum tuning: - Tactility

This child over-focuses his tactile attention on specific sensations within his environment and has difficulty 'shifting' that attention to other salient sensations. This is the child who delights in one specific tactile experience. Again, the first phase of treatment involves the creation of an adapted environment in which the child can be placed, whilst we provide appropriate stimulation designed to re-tune his sensory system. The child is then placed within that environment several times daily, - if he will allow that. Within these 're-tuning' sessions, we must first of all provide a tactile stimulus that the child will find interesting; - this will give him a point of focus for his attention. Within this controlled environment, we must now introduce several competing stimuli in order to encourage the child to 'switch' his attention. The intensity of these competing stimuli must be steadily increased until they cannot be ignored. Once attention has been successfully switched from one stimulus to another, the procedure must be repeated again and again, day after day, week after week, until the ability is reinforced and the child can accept and acknowledge several features of his tactile environment.

Over-amplification

Over-amplification: -Vision

Remember our visually over-sensitive child from the first section of the book? The child who does not like eye – contact and hates bright lights? His immature sensory processing system is having the effect of amplifying the visual stimuli being sent from the eyes to the brain. Never use flash photography near this child, or expose him to strobe lighting or any other source of intense light.

I have known situations where these sensory difficulties are not even taken into account and children who have a sensory modality, which is exceptionally sensitive, are exposed to sensory stimulation of ever - increasing intensity. One can only imagine the anguish and the horrors, which these children experience.

So, what should we do to help a child who is visually over-sensitive? How should we treat him? First, we have to create a 'controlled' visual environment. The normal visual environment, which we all inhabit, is simply too much for this child to cope with; - it overwhelms him. Where possible, this child needs a visually subdued environment. Take the mobiles out of his bedroom, take the bright pictures off his walls and remove anything from his room, which may reflect light; - this could well be keeping him awake! Do as much as is humanly possible, in a similar fashion to control the visual environment in the rest of the home and anywhere else he may go. If he ventures outside, dampen the environment as much as possible by making him wear sunglasses, indeed, if it is a bright day, think twice about exposing him to it.

I realise it may be the case that the visual environment to which the child is exposed cannot be permanently changed, it may simply not be a practicable thing to do! I propose repeated sessions within the adapted, controlled environment during each day; - if we can even temporarily successfully control the visual environment to which he is exposed, his extreme reactions and behaviour should reduce correspondingly. We are effectively attempting to re-tune the sensory filter by allowing it time to adjust and we are allowing the neurology the time it needs to recover from the overwhelming sensory battering it has been receiving.

When we feel the child is ready, we can begin to introduce visual stimulation within the subdued environment, albeit at a level and intensity, which he will tolerate. This is a matter of reducing the intensity of the stimulation to a level, with which the child is comfortable and then slowly, but surely building up the intensity to a normal level as his progress allows. This is no different than a newborn baby does for itself. – Most of the time, baby sleeps, allowing itself short bursts of interaction with the environment. Very quickly however, the environmental stimuli becomes too much for baby to handle and off he goes for another nap! Initially newborns sleep a great deal, only very gradually increasing the time they are awake and interacting with this new, over-stimulating environment, into which they have been born.

When normal sensory function has been achieved, we can begin to stimulate development as normal, assessing where the child's visual level is currently and aiming for the next level up on the visual developmental profile. It sounds easy and logical does it not? In truth, it is far from it! The practicality of even temporarily isolating a child from the normal environment is an exhausting and mentally challenging task.

Over-amplification: -Hearing

Children who experience difficulties with auditory over-amplification have a sensory tuning system, which is allowing too much of the auditory message to pass through to the processing centre in the brain, which is consequently overwhelmed. This child hates noise; - the louder the noise, the more he hates it. Even noise, which you and I would consider gentle, can be perceived as a threat by this child. He may be particularly upset by sudden noises such as a cough, or the click of a floorboard, which may elicit an over-reactive startle reflex. I have even witnessed noises such as this, elicit an epileptic reaction in children with this problem.

This child may withdraw into himself as a defence mechanism, to try to keep a world, which is too loud and threatening, at arms length. You should never make unnecessary noise around this child, indeed everything around him needs to be gentle and quiet. If speaking to him, you should whisper, he may find anything more to be too threatening to deal with.

I have seen many cases of auditory over-amplification, but I remember one child in particular, with whom it was my privilege to become involved. This little fellow had become so hyper-anxious due to his problems that he had been prescribed quite large doses of Valium (Diazepam), in an attempt to try to offset his anxiety. The problem had been worsened by the fact that his parents, out of pure love and in wanting to help him had been, for the past three years, following an eight – hour per day programme of sensory stimulation. In sensory terms, this child just needed to calm down and not be exposed to ever - increasing levels of sensory activity, which must have been a horrifying situation for him to be in. My first reaction was that this child needed some time to be left alone, to try to forget what had been happening. My first advice to the parents therefore, was to take him home, keep him and every - where else around him quiet and allow him to calm down. I advised that this philosophy should be followed until the child demonstrated that he was beginning to relax.

It took four months of doing nothing before the child began even to begin to be receptive, at which point I introduced some gentle vestibular stimulation along with sessions of gentle auditory stimulation within an adapted environment to try to clam the auditory over-amplification.

Very gradually, working alongside his parents, his GP and his physiotherapist, we were able to turn around the negative situation to which this child had been exposed. The hearing began to calm, the anxiety began to dissipate and very, very slowly we were able to reduce the level of Diazepam he was taking. It took a great deal of time and patience, but eventually the child began to make progress.

So, this story should give an insight into how we treat auditory over-amplification? Well, as I have implied, in a similar vein to treating all other sensory over-amplification problems, we need to create an 'adapted environment.' Again, we take our special room, (the quietest room in the house, well away from the noise and bustle of domestic life) and we strip it of any object, which can potentially make noise; - The room should be as silent as humanly possible. The child should be placed in the room several times a day for as long as he is happy in there. This will give him time to relax from the sensory onslaught of the real world and give his auditory system the time to start re-tuning.

Once you feel that your child is settled and content in the environment, you should begin to introduce auditory stimulus at a *very low level!* When I say 'very low level' I mean so low that you can barely hear it. Gentle, classical music is a good suggestion, particularly baroque music and Gregorian chant, which have been proven to alter the pattern of brainwaves to a more relaxed state. Be careful however, if using cassette tapes or CD's as sometimes the players can 'click' when changing CD's, tapes etc and can in a fraction of a second, undo the good work of the previous hour!

Hopefully, as your child becomes able to accept auditory stimuli at a certain level, it will be possible to very gradually increase the volume and over a prolonged period of time, bring it up to a normal level.

Over-amplification: -Tactility

The child, whose sensory system is over-amplifying the tactile sensory stimulation from the environment, is overwhelmed. The part of the cerebral cortex, which processes tactile information, is being over-excited by the thalamus. This child feels things far more keenly than you and I. The slightest touch can sometimes be amplified into an excruciating experience and consequently this child fears contact with anything, but especially contact with other people. To this child, human beings are volatile, unpredictable features of his environment, who are likely to expose him to tactile stimulation without notice. This may come in the form of being hugged, which he detests, or could involve other personal contact. He will go to extreme measures to avoid such experiences.

The first item on any agenda for treatment is yet again to create an adapted environment, into which the child can be moved for short or longer periods, as he will allow. All substances within this environment, which could come into contact with his skin, should be smooth and soft; there should be no rough textiles, no hard surfaces and definitely as little human physical contact as possible. (Except that, which is necessary to keep him safe). This is an environment where the child must feel safe and secure and an environment within which he must feel confident that no sudden tactile experience will befall him.

Hopefully, after a while, you may see that he will begin to understand that this is a safe environment for him and his enjoyment and sense of ease should become apparent. When this has been achieved, it is appropriate to begin to introduce some tactile stimulation into the situation. The stimulation must be very brief and initially at a very low level. Any sign of discomfort on the part of the child means the stimulation must be stopped immediately. It may take a while to find the correct level of stimulation with which the child is comfortable and it may seem to be at such a minute, gentle level that you do not see what possible benefit he could be deriving from it. Do not worry about this. The fact that you have your child in a situation where he is accepting any level of tactile stimulation at all is a minor miracle; - and from little acorns do mighty oaks grow!

If you find a level of tactile stimulation with which your child is comfortable, continue to give this stimulation several times a day for a prolonged period and then, when you feel he has adapted to it, increase the intensity little by little, one small step at a time and begin to give him differing kinds of tactile experiences. This process should be continued until the child is accepting of normal levels of stimulation. It cannot be stressed highly enough however that should your child show even the slightest degree of discomfort with what is happening, you should stop and even be prepared to take steps backwards.

Under-amplification

Under-amplification: - Vision

The child who is visually under-sensitive has a sensory filter, which is not allowing enough of the visual message from the eye, to be processed by the relevant part of the brain in the visual cortex. It is in fact not exciting the cortex sufficiently Many children with this problem can seem fascinated by bright lights, sometimes waving their hands in front of their faces. These children are actually self-stimulating their visual systems, - to be precise they are self - stimulating their cortex via stimulating the pupillary light reflex.

With children who are under-sensitive, treatment must aim to increase the intensity of normal environmental stimulation. This can for instance, involve stimulating the pupillary light reflex many times daily in order to stimulate the under-sensitive visual system. This child needs his visual environment to be bright and well illuminated and needs high definition contrasts in order to notice anything. Anything shown to him must be of high definition and brightly coloured. Fluorescent lights and luminous items are other good tools. Anything, which increases the intensity of the visual stimulus is good, but only within safe levels for the eyes

I will not describe the actual techniques here; for fear that some people may attempt to carry out the techniques at an intensity, which would harm the child's eyes. In fact, ***no techniques described in this book should be carried out on a child unless that child has been properly assessed by Snowdrop and the techniques have been properly taught to the individuals concerned.***

Under-amplification: - Hearing

Children with auditory under-amplification problems are children who delight in loud noises. Yes, you have guessed by now, the sensory tuning system is not allowing enough of the auditory message to be processed by the auditory cortex. You will sometimes find these children self-stimulating their sensory systems by shouting and screaming, by banging about and by generally being as noisy as possible. (Sounds like most uninjured kids doesn't it?) So, how do we treat auditory under-amplification?

The most important treatment principle is safety! It is so easy in thinking that we have to increase the intensity of stimulation, to do so without regard to safety and to do physical damage to the apparatus of the ear. ***Never attempt to carry out stimulation without first having your child properly assessed and having the benefit of qualified advice.***

Again, the general principle is that we take a room in the home and create an artificial environment, which is designed to accommodate the particular sensory difficulty with which we are dealing. In this case we would be looking to create a highly enriched auditory environment with a variety of sounds of good intensity. As the child's brain systems gradually re-tune to accommodate the auditory environment, we would bring the intensity level of the stimulation down towards normal background levels of noise. From that point, we would be attempting to stimulate auditory development normally.

Under-amplification: - Tactility

The child who experiences tactile under-amplification difficulties is simply not processing enough of the sensory message in the relevant part of the sensory cortex, which is in a state of neurobiological under-arousal. In order to address this problem, we must provide the information the brain requires at a greater intensity than normal. Within our adapted environment, we must provide wide ranging tactile experiences at a higher than normal intensity. You should find that your child will enjoy this, as at last he is able to fully appreciate sensory experiences, instead of him trying to provide those sensory experiences for himself by scratching, biting and otherwise self – stimulating himself.

Hopefully, as he becomes more able to appreciate sensory stimuli, we should begin gradually to reduce the intensity so that eventually it approaches normal, environmental levels. If treatment is successful, we can then integrate the child into a normal environmental setting.

Another reinforcement to parents here. *Do NOT try to diagnose and prescribe activities for your child from this book. Increasing the intensity of sensory stimulation to dangerous levels is very easy to do and can result in your child being damaged. It is important that your child be evaluated by Snowdrop who will then prescribe and demonstrate how to administer sensory stimulation at the appropriate, safe levels for your child. Do not do anything to your child without qualified advice!*

Internal Tuning

Internal Tuning: -Vision

This child's problem is that he cannot pay visual attention to the outside world, because the visual world, which is being created by his own sensory system is so powerful and all pervading, he cannot pay visual attention to anything else. This child is literally lost in a visual world of his own. The relevant processing areas of the cortex are being excited by the internal stimulation of their own creation. We must encourage this child to look outwards, to pay attention to a world outside of the constant background stimulation, which is being provided by his own sensory system. We do this by again, increasing the intensity of the stimulus from the outside world whilst making it explicit that this is where the stimulus is coming from.

Once the child begins to notice and to readjust his attention to the external world, we can begin to very gradually reduce the level of intensity of the stimulus and ultimately stimulate the normal development of his visual system.

Internal Tuning: - Hearing

As is the case with vision, auditory 'internal tuning' is where the sensory system creates and is tuned to its own auditory stimulus. People who suffer tinnitus will tell you how difficult it is not to be focused upon the noise inside their head. Similarly, children who have suffered brain-injuries such as those, which cause autism, can experience similar phenomena. The trick here is to achieve two things. Firstly, we must enable the child to clearly identify and separate the stimulus, which is emanating from the outside world, from the stimulus, which is being created by his own sensory system.

Secondly, we need to facilitate him tuning into the stimulus from the outside world and tuning out the stimulus, which is created from within. We achieve these two objectives by affording the child repeated opportunities, within the room we have set aside for treatment to experience noise, which is at a greater intensity than the noise created by his own sensory system.

As the child begins to respond and to tune in to this externally generated noise, he will hopefully have less of his attention focussed upon his own internally generated noises. As this ability becomes reinforced, we should be able very slowly to reduce the intensity of the externally generated noise to normal levels.

Internal Tuning: - Tactility

A child who experiences problems with tactile internal tuning has difficulty separating tactile stimuli from the outside world from tactile stimulation, which is being produced by his own sensory system. It is our task to assist him in recognising the difference and in teaching him to tune his sensory system towards the outside world. The way in which we achieve this is in very similar fashion to the child with under-amplification difficulties. We must raise the intensity of the external stimuli to a level where it exceeds the level of stimulation, which the child's own sensory system is providing. Only then will the child have the opportunity to recognise that there is in fact a sensory world outside the internally generated world of his own system.

Hopefully, as we are gradually able to re-tune his attentional systems towards external instead of internal stimuli, we may gradually bring down the intensity of the tactile stimulation we are providing towards normal levels.

Treating language and communication difficulties

Most children who suffer from brain-injury, especially the type of brain-injury, which expresses itself as autism, experience difficulties with language and communication, indeed many children possess no language at all and it is with their problems we must begin.

Many children who have not developed to the level of producing spoken language have also not developed any understanding of language, whilst some children clearly *do* understand but cannot produce language. Clearly, the more severe the brain-injury, the more likely it is the child will fall into the first category.

One possible explanation of a child's difficulties with language, which should always be investigated, is whether an auditory deficit or other sensory impairment is likely to be causing the difficulties with language development. In the absence of this problem, we move on to consider the impact of other aspects of the child's brain-injuries and their effect upon the development of language.

As I pointed out in the first section of the book, communication is not merely a useful ability to possess; it can be a matter of ***life or death***. So how do we encourage the development of language and communicative abilities in children who do not possess these skills? The first problem we encounter was discussed in the first section of the book. Because of the stress of dealing with their child's problems and the abnormal situation in which they find themselves, (allied to the possible stress of the reaction stages after the disclosure of the diagnosis of their child's problems), parents often find it difficult to interact with their child in the same way in which parents would interact with an uninjured newborn. The parents may also miss the child's early communicative signals because of the constraints, which brain-injury places on the child's abilities.

The lack of this normal interactive process ensures that baby is not provided with the opportunities he needs for the processes, which underpin language development to take place.

Let us begin at the beginning. We have already discussed the newborn baby's seemingly innate attraction for faces. This is good, because faces communicate and thus baby is already at an advantage concerning his language development, as he will naturally seek out this primary source of communication. Now think about what Mum and Dad do, when baby is scanning their face: They will move close in towards baby (Aren't they clever to know that baby's focal length is so close?) They will adopt exaggerated facial expressions (this is communication!) and they will speak to baby in a high-pitched tone (Aren't they clever to know that baby's hearing is tuned to higher-pitched sounds?). All of this in turn stimulates baby, who excitedly responds by making a noise or a movement. What the parent does next, in response to the noise, which baby has made is instinctive and absolutely critical. –

HE / SHE ATTRIBUTES MEANING TO IT!

This scenario continues each time they interact with baby and consequently, eventually the penny drops: - Baby begins to understand that sounds, bodily movement and facial expressions can have meaning and what is more, the sounds, which *he makes*, can have meaning and can influence the actions of others. This is not an overnight phenomenon, it takes weeks of interaction and the processes involved are complex.

In this way, within these communicative episodes, parent and child develop a shared focus of attention, *(intersubjectivity)*, which as we have discussed is an important factor for later language development.

So what can we do to address the problems, which are caused by the lack of these normal interactive processes between parents and brain-injured children? Well, the first thing to do is to make parents aware of the problem and how and why it has occurred. We must then make them aware of the ways in which they would normally have interacted with their child and if necessary, train them in the techniques of those interactions. In essence, we take parents back to those initial, vital interactive processes, which have been missed, in order to begin the process of language development in the child. My approach to this problem has been to develop a series of techniques, which I call *'recursive communication therapy.'*

My methodology is based upon recent research into the development of interaction between mothers and babies, which emphasises several critical features of those interactions.

These techniques require the establishment of a close relationship with the child, which is obviously best achieved by the parents. The major requirement of this approach is considerable *patience*. The adult exhibits tenderness towards the child and takes an interest in any spontaneous behaviour, responding positively and with pleasure no matter how limited the behaviour may be. When some rapport is established, the adult begins to imitate something in the child's behaviour and when the child becomes aware of what is happening, the adult turns it into a game, alternating their own response with that of the child, as a form of turn-taking. After a period of time, when this pattern of interaction has been established, the child might begin to introduce new behaviours, which is exactly what we want to happen. The child must always stay in control of events and the adult must be sensitive to this, quickly responding to signals, which indicate that the child wishes the session to stop.

It is very important that the parent is sensitive to the child's sensibilities. If during the interaction the child gives a signal of being disinterested or stressed in any way, it is most important that the interaction stops immediately and the child's wishes be respected. Remember, learning can only take place when the child is motivated to learn, we should not try to force an interaction if the child is not conducive towards the situation.

Another important point involves the issue of time: - People who have suffered brain-injuries process information much more slowly than uninjured people. It is therefore vital that within the interactive sequences, the child is allowed extra time to process the information and extra time in which to react.

There must be nothing more frustrating for a child who is constrained by brain-injury than to be attempting to interact in a turn–taking scenario with its parent, only to be given insufficient time to react to what the parent has done and to consequently miss its turn! The frustration of the situation might be enough to make the child *give up trying*!

This approach is based on research into early communicative behaviours, which has highlighted the role of joint attention between parent and child and the importance of play in establishing communication. (Josefi, O. and Ryan, V. 2004) It also makes use of Vygotsky's *zone of proximal development.*

I am aware that I have mentioned Vygotsky and the zone of proximal development several times and you are probably wondering if I am now labouring a point. The answer is that this concept is so important for our children that it is necessary for me to be as sure as I can that you have captured the general ideas behind it.

> *The zone of proximal development revisited: - This is the distance between the actual developmental level as determined by the child's independent actions and the level of potential development as determined through what the child can achieve with adult assistance, or in collaboration with more capable peers. What a child can perform today with assistance he will be able to perform tomorrow independently, thus preparing him for entry into a new and more demanding collaboration. These functions could be called the "buds," rather than the fruits of development. (Vygotsky, 1978:86-87).*

The other central features of recursive communicative therapy are:

- The distance at which interaction is initiated. – This can be particularly important with children who experience sensory tuning difficulties, who may feel threatened by close physical contact.
- The point of reference of the therapist towards the child. – This again is determined by the child's sensory problems and how tolerant they are to the presence of the person carrying out the therapy.
- Low-level involvement in child initiated play. - The parent follows what the child wants to do or to play with, and never imposes an agenda upon the child.
- The use of language to follow, rather than to control the child's behaviour. – The parents' language is always based on what the child is doing and is pitched at an appropriate level.

By following these guidelines, parents can help children to explore activities and interactions in a non-pressuring environment. The parent is also demonstrating that he / she values the child's input and actions. In this way, true turn–taking patterns are allowed to develop in a natural way and the parent is able to provide language input, which is appropriate to the child's abilities.

Recursive communicative therapy is particularly useful with children who avoid contact with others, whether they can understand and use language or not. As a result of their sensory problems, a significant feature of the difficulties faced by some children who have brain-injuries is that some of them seem to need complete control of their physical space. This means that the parent may need to keep some distance from the child. Making physical contact, moving into their space, or even looking directly at the child can take place only when the child is willing to allow it, or when the programme of sensory normalisation has made it possible. So recursive communication therapy might begin with the adult imitating in an unobtrusive way, some behaviour, which seems to please the child, followed by joining in and ultimately modelling alternative behaviours.

Not all children with brain-injuries experience such severe problems with language and some do make developmental gains in this area. So what can be done to help not only these children, but uninjured children in order to encourage maximum language development and hopefully the future development of literacy?

Quite simply, the scope of a child's vocabulary may be a determining factor in the child's future ability in literacy; indeed many studies have unearthed evidence of a link between vocabulary power and academic success. Children therefore, when they are ready for it in sensory terms, need to be exposed to as much talk as possible. If we are to widen the scope of a child's vocabulary then we must expose him to and encourage him to produce as much talk as possible. (Corson, 1988).

As children develop, they begin to be able to make a distinction between individual speech sounds and later still, they become aware of syllables. Later still, they become sensitive to the *'onset and rime'* of words. This developmental sequence continues further, with children demonstrating sensitivity towards rhyme and alliteration, which is another precursor to literacy. (Bryant & Bradley, 1996).

> *Onset and rime: - Onset and rime are used to describe units of a spoken syllable. A syllable can normally be separated into two parts: the onset, which is made up of the initial consonant or consonant mix, and the rime, which consists of the vowel and any concluding consonants.*

It seems that for pre-reading children, this developing awareness of the 'sound system' of the language is crucial and is a powerful argument for children being encouraged to use spoken language as much as possible, in order to receive sufficient practice in using the 'speech sounds' of their language. It seems there is a crucial link between the sensitivity of the child to the sound system of the language, (particularly regarding rhyme and alliteration) and later success in literacy.

All of these research findings make treatment options obvious and explicit. We need to talk to our children and to encourage them to talk to us. We need to develop their understanding of the sound system of their language, in particular their sensitivity to the onset and rime of words, the syllable structure of words and to rhyme and alliteration.

The development of verbal thought in brain-injured children

Verbal thought is something we all take for granted. As you read this page, you should be able to 'hear' a voice inside your head, which is 'reading' the words. This is called ***inner-speech***, or ***verbal thought*** and many children who suffer brain-injury do not reach the stage of cognitive development where they possess verbal thought.

Anyone who takes the time to listen to an uninjured child who is younger than the age of six will hear them giving a running commentary on their activities. Even when playing in groups they sometimes do not appear to be 'conversing' with each other, but merely engaging in what Vygotsky called ***collective monologues***. It was Vygotsky, (1996), who first postulated that this external speech is eventually internalised to become verbal thought. Verbal thought, which has developed from the child's own external speech, enables the child to develop abstract, symbolic thought, which is an essential precursor to the symbolism of written language. Vygostsky asserted that the external speech of children should be thought of as a form of action. In the same way in which non-verbal action is internalised to become non-verbal thinking, verbal action is internalised to become verbal thought. So not only are children talking in order to learn, they are also 'talking to think!'

Vygotsky argued that this internalised speech forms the basis of 'higher mental processes.' Not only does this 'inner-speech' influence thinking, it also serves the function of assisting the child in learning to read. Without verbal thought, the child loses a valuable tool in assisting the retention of word order in a text and therefore reading comprehension suffers.

Significant support for the theory that verbal thought is involved in supporting memory during reading is available. In one study, subjects who had their verbal thought repressed during reading were found to be poor in the skill of reading comprehension. Anastasiow, (1979) Baddeley and Lewis, (1997).

The question therefore is how do we foster the development of 'verbal thought' in children with brain-injuries? Well, following the principles of the early interactions between children and parents, which are described in previous sections of this book, we must first attempt to develop an understanding of language and then attempt to develop speech in our children. Once we have achieved this, our objective should be to encourage internalisation of speech; - there are specific techniques, which can achieve this:-

- Encourage children to use rehearsal. – This is a good technique to help develop verbal memory. When you wish to remember a telephone number for instance and you do not have a pen to hand, what do you do? You continually repeat the number over and over to yourself. This reinforces it in verbal memory and means you are unlikely to forget. If we can learn to rehearse, so can our children.

- Provide a running commentary of their activities for them and encourage them to talk along with you. This will teach them how to develop external language and help to develop verbal memory. When this skill is developed, it is time to encourage your child to say the words in his head; - this will help to facilitate the development of verbal thought.

Although Vygotsky's ideas receive significant support from educational research and this is undoubtedly an important route towards developing literacy in children, it is not the ***only*** way in which the child can develop literacy skills. One thing I have learned as far as developmental theories are concerned is that no-one has a monopoly on the truth. With this in mind, we now examine another method of teaching literacy abilities using a totally different method.

Top-down processes

There appear to be two pathways, which the brain uses for reading, which we will come across in more detail later when we discuss dyslexia. The first is the ***phonological*** pathway, which corresponds to all of the evidence we have just discussed concerning the development of verbal thought. The second is the sight – recognition pathway, where children recognise words simply by seeing them. This second pathway can also be used to teach children to read.

Whilst it is true that developing an understanding and competence in the sound system of the language we use, is valuable in helping children to learn to read, it has also been proven that the process can also work the other way. Learning to read can be helpful in helping children to grasp the sound system of their language. (Bruce, 1964). The idea that letter identification must be complete prior to word identification taking place, appears to be incorrect. The *word superiority effect*, demonstrates that word recognition processes can and do influence identification of the letters of a word. (Eysenk & Keane, 1997).

The response to this in terms of treatment techniques is that as soon as the child's visual pathway is capable of seeing a written word, we should institute a reading schedule.

A reading schedule can be devised, using these *top down* processes to help teach children who have brain-injuries to read. Any reading session carried out with your child should be enjoyable and connected with rewards or other pleasant activities, which will help to motivate your child towards the activity.

A reading session, (or any other form of developmental stimulation), should never be carried out if the child is not in the mood for it. Remember learning only occurs when the child is motivated towards it! Do not worry if you feel your child is not learning to read the words, be assured that the brain, if presented with information in the correct manner is a wonderful learning machine and it will be soaking up the information.

So, we have two processes by which brain-injured children can be taught to read, the question is, which one should we use? - The answer? **Both!** There can be no room for the academic egoism displayed in nailing our flag to the mast of a particular theory and refusing to consider anything, which falls outside the boundaries of that theory. If there is evidence to support the use of a technique, (which there clearly is, in the case of both 'bottom up' and 'top down' reading processes), then we use it. We use every tool at our disposal to fight this enemy called brain-injury, which possesses our children.

Having briefly discussed the treatment of language and communication difficulties, it is now appropriate to discuss the treatment of perhaps the most obvious sign of disability to the outside world: - that of mobility problems.

Treating mobility problems

Children, who suffer from brain-injury, including some who are diagnosed as autistic, invariably endure mobility problems of varying degrees of severity. What can be done to improve their performance in this critical area of development? There are two things to which I wish to draw to your attention concerning mobility problems in our children. - The first is ***opportunity*** whilst the second concerns the word ***success.***

- ***Opportunity***: - When a well, uninjured child progresses beyond the initial babyhood period, he begins to spend more and more time on the floor. Gessel, (1925), referred to the floor as the child's 'developmental workshop.' It is through spending time on the floor that the child learns to roll, crawl, sit and eventually to walk and run. However, as brain-injured children progress past the initial babyhood period, they ***do not*** spend increasing amounts of time on the floor. It is not considered fair to leave a brain-injured child to his own devices on the floor, instead we place him in a chair, or a standing frame and consequently his 'developmental workshop' is withdrawn from him. Therefore, we further compound the effects of the brain-injury on his mobility.

I completely understand the rationale for this different treatment of brain-injured children, - it is borne out of a caring attitude; - out of a wish to protect and care for children; - it seems the natural approach to take. In fact, this philosophy towards brain-injured children and the floor is exacerbating their mobility problems, with children being deprived of the very environment where development can take place.

- ***Success:*** - The point I make here will initially seem to contradict the point I have just made concerning the floor, but stick with me, eventually I will make myself understood! A child with brain-injury, if placed on the floor and asked to move will usually fail. His musculature is weak and badly coordinated and the further effect of gravity upon this compromised situation acts to guarantee failure. So, our child fails; he is accustomed to failure in everything else he attempts, so further failure here is no surprise. With failure comes lack of motivation and an unwillingness to even attempt to achieve.

As I say, I am aware that it seems as though I am contradicting myself because in the previous paragraph, I was extolling the virtues of the floor and now I am saying that if you place your child on the floor, you are creating failure.

The simple answer is we have to adapt the environment in order to create success; - much in the same way that we adapted the sensory environment to create successful sensory development. So how do we achieve this? The answer lies in creating a situation where we combat the effects of gravity, (indeed, where we use gravity to our advantage) and where our child learns that movement is not associated with failure, it is associated with success and with enjoyment. To this end, we construct a *ramp*, which is a flat 'door sized' piece of wood or other solid material.

The construction does not need to be complex; indeed a flat door placed at the appropriate angle would suffice. The ramp needs to be a minimum of seven feet, (2.2 metres) in length, depending on the height of your child, and of sufficient width so that he fits comfortably onto it. The surface of the ramp should preferably be a smooth material and should be padded, so as to prevent injury if a child should let his head drop. The plane should preferably have sides attached to it so that the child cannot roll off and fall; - failing this, an adult should stand at each side of the plane so that there is no chance of injury in this way. This technique is used in various forms at other clinics around the world and was first utilised by Doman in the U.S.

The ramp should be placed at an angle so that when the child makes a limb movement, however slight the movement may be, he moves forward just a few inches. This will have the effect of teaching him to associate his limb movements, with forward movement. It will also teach him that movement can be easy and enjoyable. This will create a situation where your child is *succeeding* and his success will create ***motivation!*** Over time, as your child begins to move more easily, the angle of the ramp should be very gradually reduced so that it is closer to horizontal, with the eventual aim being that your child is able to crawl on the plane horizontally. Your child is now strong enough, coordinated enough and motivated enough to be placed on the floor!

When a child is ready for the floor, it would be easy to leave him there for prolonged periods in order to maximise his opportunities for developmental progress; - this would be a mistake and would most likely lead to the situation which existed before, which was one of failure and lack of motivation. Floor-time should be gradually increased, as and when the child is ready to accept it. Floor time should incorporate games and pleasurable situations, which the child will then associate with it. At the first sign that your child is becoming frustrated or disinclined to participate, then you should abide by his wishes. Where this situation exists, it may be necessary to revert to using the wooden ramp for a short period to rebuild success and confidence. Remember, only ***success breeds motivation*** and motivation then breeds success!

When your child is crawling, we will wish to encourage passage through the normal developmental pathway of mobility, which will lead to him crawling on just his hands and knees. In order to do this he will need more sophisticated control of balance and coordination. This is achieved through a programme of ***vestibular stimulation.***

The vestibular system in the brain is responsible for much more than the mere ability to crawl on hands and knees, stand erect, maintain balance and move without falling. It organises incoming information from the eyes, inner ear, muscles and joints, fingertips and palms of the hands, the soles of the feet, and gravity receptors on the skin. It also helps to regulate heart rate and blood pressure, muscle tone, limb position and level of arousal.

> *The vestibular system, or balance system, is the sensory system that provides the dominant input about our movement and orientation in space.*

Problems in the vestibular system can cause the type of hyper-anxiety we see in many children with brain-injury. Vestibular system difficulties cause abnormalities in muscle tone, (muscle stiffness), difficulty passing stools, teeth grinding and slobbering, (drooling), etc. Much of these are explicable by its influence on the ***autonomic nervous system.*** I have found that exercises, which provide stimulation to the vestibular system to be effective in reducing many of these vestibular problems.

> *The autonomic nervous system is the part of the nervous system that is not under conscious control. It is commonly divided into two usually antagonistic subsystems: the sympathetic and parasympathetic nervous systems, and involves controlling the equilibrium of organs and basic physiological functions.*

Vision is an important unit of the vestibular system. Twenty percent of visual brain cells act in response to vestibular stimulation. People who have suffered injury to the vestibular organs in the inner ear can be taught to utilise visual information to maintain balance. However, if the visual information is disengaged (e.g. like when we are in a darkened room, or there is some sort of optical illusion), the individual will feel disturbed, as though they were falling.

The auditory system is also highly involved in the function of the vestibular system. The vestibular and auditory nerves link in the auditory canal to form the eighth cranial nerve of the brain. Anything, which interrupts auditory information, may also affect vestibular functions. So, you can clearly see how important a schedule of vestibular stimulation is to the child with mobility problems.

The vestibular system has also been shown to influence the sleep /wake cycle. Many people with vestibular problems have a poor sleeping pattern. Does this ring any bells concerning our children?

Upon assessment, your child may be prescribed a range of vestibular activities, depending upon the problems he displays: - These may range from spinning, swinging, rolling, and rocking, through to simple visual activities.

The importance of physiotherapy

There are some alternative treatment regimes, which hold that physiotherapy is of no use to the child with brain-injury. Some of them will not have anything to do with physiotherapy at all. I do not subscribe to this point of view. The role of the physiotherapist is **vital** in the management and treatment of brain-injury. Some children with autism will not require physiotherapy, but some who have symptoms mixed with cerebral palsy for instance, will. This is why I consider the most important person for me to work with and to coordinate treatment with, is the child's physiotherapist. The goals of physiotherapy are simple: - The prevention of postural deformity and the facilitation of normal patterns of movement. These goals are achieved using principles, which include the following.-

The inhibition of normal reflex patterns, which may not have been naturally suppressed. As children with brain-injuries grow older, the activity of their central nervous systems can fall into well-worn patterns. A good example of this is seen in the child with very stiff muscle tone. In this case, the pattern of activity, which the nervous system has adopted, has the effect of increasing the child's muscle tone. Over the years, as the pattern of neurological activity becomes well worn, the rigidity of the musculature increases. It is thought that this occurs because, as a consequence of the brain-injury, there are only a limited number of **synaptic chains** available to register a response to any stimulus. (Bobath, 1980).

Synaptic chain: - Groups of brain cells, which are interconnected and usually communicate in a serial manner, the previous groups communicating with the next groups of neurons in the chain

---

It is the function of treatment, if successful, to attempt to increase the number of synaptic chains (neural networks), which are available to register a response. Unfortunately, physiotherapy cannot achieve this, through no fault of its own; - physiotherapy simply is not able to improve developmental prospects.

However, what physiotherapy can do is serve to inhibit these abnormal motor responses, (the increased muscle tone), by means of handling the child in such a way as to prevent a state of increased muscle tone occurring. In inhibiting these abnormal patterns of movement, the opportunity is then created for more normal patterns of movement to occur.

So, although physiotherapy is of no use as a tool, which can stimulate child development, as proven by research carried out by Scherzer (1976) and Wright and Nicholson, (1973), it can act to prevent the negative aspects of brain-injury from overwhelming the child's body. This can then help to facilitate a programme of developmental stimulation having maximum opportunity for success.

Learning difficulties

How the brain learns. – Plasticity in action.

In order to learn, we first have to perceive; - In order to perceive, whether by vision, hearing, touch, taste or smell, our sensory systems have to be functioning correctly. Therefore, learning is subservient to perception, which in turn is subservient to development of the sensory system. Therefore, the first goal in any learning programme with a child who has brain-injuries is normalisation of the sensory channels. We have previously discussed this.

When we learn, our brains respond to new information and as such, they must necessarily adapt their functioning. They achieve this through ***synaptic plasticity***, - the forming of new connections in the brain, or the strengthening of existing ones.

Yang & Maunsell, (2004), in their work with monkeys were clearly able to demonstrate that learning involved changes in synaptic connections in the cortex, which established new neural connections. Again, we are back to the concept of ***neural networks***, - groups of brain cells operating together as a network to encode (learn) new information. In order to represent one piece of learned information, the network will connect together in a specific pattern, and in order to represent another piece of learned information, the same network might slightly alter its pattern of connections. This is the way in which brains learn and remember. – A memory is merely a representational pattern of connections!

There are many types of learning of which human beings are capable, which will create such connective networks in the brain. One of these, probably the simplest, is the cause and effect learning of ***classical*** and ***operant conditioning***.

Classical Conditioning is the style of learning made notorious by Pavlov's research with dogs. The general idea of the research was this: Pavlov offered food to dogs, and calculated their salivary reaction (how much they drooled). He then began sounding a bell immediately prior to offering the food. Initially, the dogs did not start to salivate until the food was offered. However, after a short time they began salivating when they heard the sound of the bell. They had learned to connect the sound of the bell with the offering of the food. In as much as their immediate physiological reactions were concerned, the ringing of the bell corresponded to the offering of the food; - It had become a ***conditioned stimulus.***

Classical conditioning however, due to its haphazard nature is difficult to use as a teaching method with children, my only reason for introducing it is to move on to another form of basic cause and effect, associative learning, that of ***operant conditioning.***

Whereas classical conditioning forms an association between two stimuli, (the bell and the food), operant conditioning forms an association between a behaviour and a reward, (or punishment). The reward ***reinforces*** the desired behaviour. – In this way we are able to ***shape*** (encourage), desired behaviours with a reward and to eliminate undesirable behaviours with negative reinforcement. It is basic cause and effect learning, which it is sometimes necessary to use with children whose brain-injuries are so severe that all other approaches to learning would be inappropriate. If we look at our strategies for teaching children as a ladder, then this is the first rung on that ladder; - the very basic first steps to learning.

With severely brain-injured children, rewards have to be instantaneous, or plainly linked to the behaviour. With children who understand some spoken language, we can make clear the relationship between the reward and the behaviour, even if they are removed in time. For instance, you might tell a child that you will give him a treat because he helped you to do something. With children who possess no understanding of spoken language, the relationship between the reward and the behaviour cannot be made explicit in this way. For these children, the reward has to be instantaneous.

As I say, this type of associative learning is the very first, simplistic step on the ladder to success. Fortunately, most brain-injured children are in advance of this level of learning and can be approached with more sophisticated techniques. However, there are some children, who have more profound brain-injuries, where these techniques can be useful.

Children who suffer from brain-injury invariably have some degree of learning difficulty. These difficulties vary in their nature from individual to individual; - they also vary in severity. I differentiate between children in what I term ***organic learning difficulty*** or ***environmental learning difficulty.***

> *Organic learning difficulties are caused directly by the brain-injury, by direct damage to cognitive structures of the brain, which cannot then operate correctly.*
>
> *In environmental learning difficulties, the cognitive structures may be intact, but the brain-injury is preventing learning from taking place because it is causing sensory difficulties, or other problems, which prevent the child from interacting with the environment.*

Of course, some children ***appear*** to have greater learning difficulties than is actually the case, simply because their brain-injuries prevent them expressing the intelligence they do posses.

Many of the problems our children display as far as learning difficulties are concerned, involves the ***speed of processing*** information. The brain is an information-processing machine, where networks of brain cells all happily and speedily communicate with one another. What then happens when suddenly one brain cell in the network is lost through injury? The remaining network of cells may still be able to process information, but without the full complement of cells in the network, the task is more arduous and therefore much slower. In time and with practice and proper stimulation, the network may become more efficient; - it may even be able to improve its performance by recruiting an additional cell to the network in order to restore the full complement. Now apply this scene to millions of networks in the brain, which are affected by oxygen starvation and which consequently lose cells and thus lose operational efficiency. Can you see how the whole process of learning and development can be slowed down?

Other major problems our children face are with memory and attention, which are vital functions needed to support learning; - if we cannot pay attention to information then our chances of learning are remote because we simply will not have access to the information, which should have been learned and therefore will be unable to pass that information to memory. If memory is affected then we cannot remember all the information, which has been presented to us. This obviously impairs the ability to learn!

Many people think abilities such as memory and attention are built into us from birth, but this is not the case at all; - they are learned! It seems the brain centre responsible for the development of these abilities is a structure in the ***limbic system*** known as the ***hippocampus***. We have already discussed how we learn to develop our attention as young babies through the early interactions with our parents, developing what we termed 'intersubjectivity.' We have also discussed techniques whereby we can stimulate the development of those attentional skills in brain-injured children by taking them back through those early interactive episodes.

In a similar vein, we also learn memory functions. This is useful, because logic dictates that if we learn how to do something then by definition we must be able to find a way in which it can be ***taught***! I have previously used the example concerning memorising a telephone number, which is unfamiliar. What strategy do you learn to employ in order to achieve this? The solution is to keep repeating the number over and over again until it is ingrained in memory? By 'rehearsing' the number verbally, aloud in this way, you are constantly refreshing it in auditory memory. (Auditory memory usually fades after about one and a half seconds). This gives the networks of brain cells the extra time they need to process the information and to embed the information within the network. (To memorise it). This is just one simple technique by which our children's memory systems can be developed. For children who experience difficulties in the area of language, whilst we are working on developing those language and communicative abilities, it may be necessary for us to externalise techniques such as rehearsal by simply doing it for them; - by repeatedly presenting the information to be memorised and learned. There are also many games, which can be used to help develop these abilities.

Therefore, in order for us to combat the learning difficulties, which are faced by our children, it is important for us to understand how children learn. One of the great names of child development is Jean Piaget. I have already highlighted that he proposed that children's development proceeds in 'stages.' As we have discussed, according to Piaget, a child cannot complete developmental tasks, which are in the next higher stage until his brain has matured beyond the stage at which it is currently operating. He proposed that maturation (simply growing up) plays a key role in the development of a child's cognitive abilities. According to this view, children are incapable of achieving tasks, which are currently above their developmental level until their brains have matured in accordance with that level. This view also holds that development through maturation of the brain, is required before learning can take place. This view of development leaves little hope for brain-injured children whose brains' will in some cases never 'mature' to higher stages of development; - Thankfully, it is incorrect!

Lev Vygotsky on the other hand, did not subscribe to Piaget's view. Vygotsky proposed that children's learning is social in nature; he asserted that all learning first takes place on a level, which is external to the child, in social situations. As learning takes place, with assistance from more skilled others, (in this case the parents), abilities, which the child could not initially perform alone are *internalised* to become part of the repertoire of skills of the child. In this way, learning leads development. This view gives great hope for brain-injured children. All we need to do is to determine the child's actual developmental level, - determine their 'zone of proximal development' and design activities which can help them achieve the next, higher developmental abilities. (It might be easier if you think of the ZPD as the 'zone of *next* development. – It is the developmental abilities which lie just out of the child's reach, but which he can achieve with appropriate training and instruction)

Let us consider what Vygotsky proposed in a little more detail, as it underpins my approach towards combating your child's learning difficulties. Lev Vygotsky was a Russian psychologist who was born in 1896 and died at thirty-eight years of age. His views on child development are only just beginning to become common currency in the Western world where the education system and child developmental viewpoint has been dominated by the views of Jean Piaget. Vygotsky concentrated his research upon the way in which children learn in natural situations. As the developmental level of most of our children will be at pre-school levels, we shall begin there.

In the pre-school years, children learn and parents teach (without even being aware of the fact that they are doing so), in the relaxed setting of the home and wider community, within interactions where more skilled members of family, friends and community assist in guiding their learning. This way of learning is natural and uncontrived. It involves the use of the *sociocultural tool of language* as a means of developing a social mode of thinking.

> *The sociocultural tool of language: - By this, I simply mean that in such a learning situation it is spoken language, which facilitates the interaction. This spoken language is 'external' (out in the open) and helps to make the thought processes of both parent and child available to the other. In this way the child can utilise the parent's language in order to guide his learning and the parent can utilise the child's language to understand the difficulties he may be facing and to adjust the level of assistance accordingly. In this way language becomes a social mode of thought between parent and child.*

There is evidence that this sort of language use and the consequent development of vocabulary power in young children leads to later scholastic advancement. - In this sense, learning occurs through talking! It is recognised that this type of social, learning situation encourages pre-school children to instigate interactions, ask questions and seek information eagerly. They consequently become active in their own learning and are able to utilise other people in their environment to help them. (Corson, 1988; Wood, 1986; Mercer, 1995). If this is the natural way in which children learn, then this is the way to pursue learning in brain-injured children.

Let's take an even closer look! Bruner coined the term ***scaffolding*** to define just this type of supported learning, which has been observed in children's homes; - learning which takes place through what Vygotsky termed the ***zone of proximal development*** (Wood et al, 1976). This is a situation where the learning, which is being supported through the guidance and support of a more skilled 'other' (the support given by the parent being the scaffold), is slightly in advance of the child's current capability to carry on successfully alone. As the child's ability to carry out a task increases, the verbal and practical support and guidance given is gradually decreased, leading to a situation where the new ability becomes ***internalised*** as part of the child's developmental capability.

The concept of scaffolding was later, further refined by Rogoff, who described such supported learning as a process of ***guided participation***. The concept of 'guided participation,' emphasises not only the cognitive support given by the adult, but also acknowledges the active participation of the child in the learning process. (Rogoff et al, 1993).

There is evidence to illustrate the success of guided participation between mothers and children and although this is mainly gained from studies of mothers and children in middle class European and North American culture, there is also support that the use of guided participation is cross-cultural. (Rogoff, 1990).

Research into the areas of conceptual development, object exploration, puzzle construction and memory tasks, all demonstrate that when the dyad of mother and child colluded in these tasks, utilising a 'guided participation' approach, children advanced their skills in these areas. (Deloach, 1983; Rogoff et al, 1984; Bornstein, 1988).

A good example is provided by one study in particular, which demonstrates that the dynamic and helpful participation of an adult in children's investigation of new objects, led to more exploration by three to seven – year – olds, than did the straightforward company of an adult. (Henderson, 1984a, 1984b,).

These results indicate that this is indeed one way in which children learn, but let us take a closer look at the 'zone of proximal development,' and how new skills are attained through it until they become a part of the child's actual developmental abilities. Consider a child who is trying to learn a new skill, which he is unable to complete alone. He needs the assistance of an adult and he is totally dependent upon that adult in order to complete the task. He is just beginning his journey through the zone of proximal development, which has four stages as follows.

- **Stage 1**: At this early stage in the learning process, a more skilled adult provides assistance in the new task. The child is completely dependent upon the guidance (the scaffold), which is provided by the adult and is unable to undertake the task alone. At this point, the ability to complete the task is *external* to the child; - it is within the social interaction between the child and the adult.

- **Stage 2:** Because of the guidance, which has been provided by the parent within the social learning situation, the child is able mostly to complete the task alone. In other words, the scaffolding of guidance has been gradually removed as the child's ability to carry out the task increases. At the end of this stage, the child is able to complete the task alone. At this point, the ability to complete the task is no longer external, within the social interaction between child and adult, but has been *internalised* by the child. The ability needed to perform the task has gone from being within the developmental grasp of the child, to actually being part of his developmental abilities.

- **Stage 3:** Automatisation through practice. Although the child has internalised the abilities required to complete the task, they are still not firmly embedded. He may still have to use his full concentration to perform the task, even utilising language to talk himself through the process. In this way, through language which is first spoken externally and which is later spoken internally, as ***'inner speech,'*** or ***'verbal thought'*** he is providing his own guidance or scaffolding. With sufficient practice, the task is gradually performed more fluently and he has to concentrate less and less in order to complete the task, until eventually, it becomes automatic.

- **Stage 4:** De-automatisation; recursiveness through previous three stages. This happens sometimes, if the learned task is not carried out for some time, or it has not been automatised through sufficient practice. We begin again at stage 1. (Tharp & Gallimore, 1988)

The way in which children with brain-injury learn is no different to the way uninjured children learn. They are both children! The learning of one is merely obstructed by the barriers constructed by the brain-injury. We have to find a way of overcoming those barriers and the way to do this is to break a task down into subgroups of much simpler tasks, which are achievable. For instance, if you were trying to teach your child how to feed himself, would you just give him the spoon and expect him to copy you, - to follow your lead? With an uninjured child, this would be possible, but to a child with brain-injuries, this task may be as difficult to him as flying a space shuttle would be to us. We must break the task down into a series of achievable sub-tasks. The first sub-task might be learning to hold the spoon correctly, which eventually would lead on to the second sub-task of dipping the spoon into the food. The third sub-task would then be to bring the spoonful of food back to the mouth to be eaten. Each of these sub-tasks would be learned by using the principles of the four stages of the zone of proximal development.

It might not necessarily be a practical task like learning to feed oneself that could be treated in this way; many other developmental tasks can be treated in a similar manner. No matter how severe the child's problems are, he is still capable of learning if the material is presented at the appropriate level and in an appropriate format. We just have to know where to start and how fast to progress.

Sleep

As promised, a small section dedicated to what I feel is one of the most frustrating problems faced by brain-injured children and their families. As the father of a child who for sixteen years, hardly ever slept, I have experienced the mind numbing tiredness and anxiety this situation produces. I have also witnessed its effect upon my child and other children and families in similar situations. The daily stress of taking care of a child with brain-injuries is compounded because of lack of sleep. So what is sleep and why is it so important?

There is an argument as to whether sleep is an altered state of consciousness or a behaviour. – The argument is academic and irrelevant if you are being deprived of it! The assertion that it is a behaviour may appear a little bizarre, because we usually visualise a behaviour as something, which entails movement. However, apart from the rapid eye movements, which are associated with a particular stage, sleep is not characterised by movement. Because we recall very little about our sleep, apart from the occasional dream, most people have a tendency to view sleep more as an 'altered state of consciousness.'

There are several structures in the brain, which are important for the regulation of sleep; they are the **hypothalamus, suprachiasmatic nuclei, reticular activating formation and pons.**

Most people are aware that there are different stages of sleep and will at least have heard sleep broken into two categories, **REM sleep** and **non – REM sleep**. There are actually five stages of sleep, incorporating REM sleep and four stages of non REM sleep. Let's take a look at the most well known phase of sleep first, REM sleep (which has a great rock band named after it!). During REM sleep, we dream and the eyes move rapidly, (hence the term, REM – rapid eye movement). This type of sleep occurs for a few minutes at approximately 90 - minute intervals during the night, which means that throughout a typical night, we dream several times. The rest of the time, we are travelling through the other, non-REM stages, with stage 4 sleep being the deepest level. During

REM sleep, our body musculature is disconnected from the neural centres, which control it. This happens for our own safety, so that we do not 'act out' our dreams. In fact, physiological studies have shown that a person actually becomes paralysed during REM sleep. (Carlson. 2007).

The ninety - minute period between periods of REM sleep seems to be important in general terms as it transfers to our everyday activities. It was Kleitman (1961), observing babies who were fed on demand, who noticed a ninety - minute cycle appearing. Later studies discovered ninety - minute periods of rest and activity in adults too, for such activities as eating, drinking, stomach motility, and ability to pay attention. (Kleitman, 1982).

These findings are significant when we consider our children. In uninjured children, it is the demand of the social environment, which suppresses this ninety - minute clock with regards to sleep. The pattern of life in the Western industrialised countries means that most people need to work all day and sleep all night. – This pattern forces us and our children to override the internal clock and within a relatively short space of time of being born, they are usually sleeping through the night. In some non - industrialised countries, in Africa for example, there are tribes whose parents carry their children around with them whilst they carry out their daily activities. Regarding sleeping, these children and parents have a tendency to fall into the ninety-minute pattern during day and night.

When a child has suffered brain-injury, he may not have the sophisticated level of neurological organisation to override this ninety - minute cycle and to comply with our pattern of life. Therefore, from our point of view, his sleeping pattern is chaotic and disturbed, as is his pattern of activity during the day. Actually, his sleeping pattern is natural, - ours is not!

It seems that we do not need to sleep in order to overcome the trials of our physical workload during the day. There have been studies of sleep deprivation with human subjects where it has been conclusively proven that sleep is not needed to keep the body functioning normally. (Horne, 1978). However, if we are deprived of sleep, it does seem to affect our cognitive abilities, so it seems that sleep is primarily needed by the brain, not the body.

So, apart from the ninety – minute biological rhythm, to which many of our children appear to be naturally complying, - why do so many of our children experience sleeping difficulties? There are many other reasons.

There are many sleeping problems our children can display depending upon the nature and extent of their brain-injuries. **Sleep apnea** is a problem for many adults, but it can also be a problem for brain-injured children. Sleep apnea is where the person stops breathing during sleep. The brain then causes the person to wake suddenly and breathing restarts. They literally have difficulty in sleeping and breathing simultaneously. Macey et al (2002) have highlighted that this problem can occur because of early brain damage and the way in which the developing, injured brain 'wires itself.' It can also be caused by a build up of carbon dioxide in the bloodstream during sleep and the child (or adult) wakes gasping for air (Carlson, 2007). The poor development of the breathing pattern of children with brain-injuries can only exacerbate this situation. Fortunately, it is only a small minority of parents who report this problem with their children.

Another problem, which concerns sleeping is **narcolepsy**, which is caused by the brain's inability to adequately regulate wake / sleep cycles. In this condition, the child will suddenly fall asleep, regardless of what is happening around him. The sleeping attack will last only a few minutes and he will wake again. This problem seems to emanate in a part of the brain called the **hypothalamus** and involves a failure in the processing of a special class of neurotransmitters called **hypocretins.** - Neurotransmitters are special chemicals, which brain cells produce to communicate with each other and to regulate biological processes. The neurons that produce hypocretins are active during wakefulness, and research suggests that they keep the brain systems needed for wakefulness from shutting down unexpectedly, which is exactly what does occur in narcolepsy.

Related to narcolepsy is **cataplexy**, which involves a sudden loss of muscle tone, so that the person abruptly falls to the floor. This can easily be mistaken for a type of epileptic activity known as **clonic** seizures. As I have discussed, one of the primary features of REM sleep is muscle paralysis, - a disconnection of the muscles from the control centres in the brain. In cataplexy, this disconnection from neurological control occurs at unusual and unpredictable times and the child simply collapses.

There are two further categories of sleeping difficulty to which, due to their brain-injuries, our children can be assigned. – Firstly, there are children who **cannot sleep**; - these children have to push themselves to a state of physical exhaustion before they can sleep and will stay awake for days. I once knew a child who managed to stay solidly awake, and to consequently keep his family awake for eleven days and nights! His poor parents were exhausted!

Secondly, there are children who can sleep, but ***do not do so at the correct time.*** (our ninety – minute children also fall into this category) Their sleeping pattern (circadian rhythm) is not operating correctly

Looking at the first category of children, those who cannot sleep, their difficulties may have various causes. First, we need to consider that the cause may be sensory in nature. It is very difficult for children who have sensory over-amplification difficulties to sleep. The slightest noise, such as a creaking floorboard, visual stimulus such as a mobile in the room, or tactile stimulus such as the sheets rubbing against the skin, will be sufficient to guarantee lack of sleep for the child who suffers from this type of sensory difficulty. As we have hinted, children with this type of problem will sometimes stay awake for days, sleeping only when pushed beyond the point of physical exhaustion.

What can be done to improve the prospects for sleep? Well, children who visually over-amplify should have a bedroom, which is free of visual clutter. The room should be as dark as possible with all mobiles and posters removed; in short, the room needs a dampened visual environment. The child who suffers from auditory over-amplification difficulties is a different kettle of fish however. It would be difficult to supply an auditory environment, which was absolutely quiet. This child hears the slightest noise, a floorboard creaking, the tap running in the kitchen; he is so sensitive he almost hears you think! - This child hears *everything* no matter where it is in the house, no matter how quiet you think it is. The only thing to do is to mask the noises, which keep him awake. This can be done by keeping a constant gentle background noise in his room. As we previously discussed, it is best not to use CD's or tape recorders, which are likely to click when the CD or tape has finished; - the most effective way is to leave a radio on in his room all night, just at a gentle background level. This should be enough to provide a mask for any other noise, which may occur.

The child who over-amplifies in tactile terms, hates bedtime. The rubbing of sheets against the skin, the cool air of the bedroom; - all guarantee a level of over-stimulation, which serves to guarantee lack of sleep. The first priority with this child is to keep the bedroom warm enough so that you do not have to cover the child with sheets or a quilt. Ensure the environment is free of anything, which may accidentally come into contact with the child. This may improve the situation.

The second cause of sleeping problems involves the brain in other ways. Sleeping patterns are influenced by the patterns of light and dark around us.

Light hits the retina and this message is passed to a set of brain cells called the ***suprachiasmatic nuclei.*** The suprachiasmatic centre is one of the body's major biological clocks. It not only regulates hormones related to the day/night cycle and therefore our sleeping cycle, but also orchestrates the activities of many other internal clocks. In numerous experiments, it has been shown that, when the SCN is not innervated, the human body clocks free-run; they set their own time. It has also been shown in studies of sleep disturbance that among the visually impaired there is a higher incidence of sleep impairment. Due to these visual impairments, the suprachiasmatic cells are not receiving enough stimulation from light in the environment. (many of our children are visually impaired to some degree. This would certainly apply particularly to children who suffer visual under-amplification difficulties).

Therefore, we now have reasons why many of our children have chaotic sleep patterns, the question is; - how can we address this?

There are many ways in which the sleep cycle can be influenced. Firstly, many of our children, because of their physical disabilities do not get the same amount of exercise as their well compatriots. Moderate amounts of exercise have been shown to positively enhance the sleep pattern. A programme of visual stimulation at the correct time of day can also stimulate the suprachiasmatic nuclei into influencing the sleep / wake cycle, as can simply allowing our children to sit in sunlight at the right time of day. There are also dietary methods, which can positively influence sleep. The right foods, at the right time of day, can help to positively influence sleep; - For instance, a small amount of banana, which contains ***tryptophan***, a few hours before bedtime, can help the onset of sleep. (Tryptophan is a precursor of serotonin, which we need to produce to bring on sleep).

If all these techniques fail, as they sometimes do, you could consult your doctor concerning a prescription of 'melatonin' a naturally occurring hormone, which promotes sleep, whose production in the brain is stimulated by a reduction in the amount of light in the evening. (***You must consult your child's doctor before administering any medication to your child***).

As an absolute last resort, in the most severe cases, your doctor can advise you on the use of more powerful medications, which may ensure a nights sleep. Beware though, these medications, such as chloral hydrate, or benzodiazepines are very strong and addictive. They can also promote side effects and 'hangover' effects.

Ultimately, sleep may always be a problem. I have known children who have taken enough medication at night to knock out an elephant and still they have stayed awake!

Practical Issues

Distorted neurological development

A while ago, one child with whom I was involved, who suffered from the type of brain-injury which increased the stiffness of his limbs, suddenly developed a marked increase in muscular rigidity. This was despite the ongoing developmental activities designed to help ameliorate such difficulties. The rigidity in his muscles and limbs was so extreme as to make him difficult to handle. This made us very cautious about carrying out aspects of his programme, lest the child was hurt.

Through research and previous experience, it became apparent what was happening with this child. As any baby develops, new neurological structures are brought 'on–line' in order to support the development of new abilities. This bringing 'on–line' of new structures was what was occurring in the case of this particular child, however the structures which the brain was attempting to utilise were badly damaged by the brain–injuries he had suffered at birth. Consequently, the structures, which the child's brain was attempting to utilise, were causing the increase in rigidity; - instead of behaving normally, the injury was causing them to behave in a distorted manner, creating a further distortion in overall brain function. Therefore, you can easily understand that as a child with brain-injuries ages, his difficulties also evolve and change. So, if your child's problems suddenly change, this could be a possible reason.

Education: - Statements of special educational needs.

This part is probably only relevant for parents in the UK, but I know that at least in the US, many parents do not send their children to school, preferring to educate them at home, so there may be points of interest here after all.

When my own son was two years old, we received a letter from a doctor who apparently 'specialised' in child development. She wanted to discuss providing our son with a ***statement of special educational needs***. Initially, we did not know what she meant by this term, but we soon discovered what it was and how strange it was that someone should want to begin this process with a child who was at the time only two years of age, - well below school age.

First, to all parents currently dealing with this situation, allow me to explain exactly what a 'statement of special educational needs' actually is. When used correctly, it is a method of recognising that a child who is currently attending a school has needs, which cannot be met by that school without additional resources, including provision for funding of those resources. I have nothing against this; it is a way of guaranteeing that schools have the ability to deal with the needs of all their pupils. It is also a method by which the exact needs of the pupil are laid out clearly for all to see. This is good!

There are however, circumstances where I would disagree that to educationally statement a child is in the best interests of that child and the family. Firstly, it is a mechanism, which can be used as a means of coercion by professionals, who wish their views to prevail over the wishes of the family. This 'coercive persuasion' may range from a professional desire to see a particular course of action taken in the classroom, to attempting to convince parents to send their child to school, when they do not wish to. It is important for parents to know that ***they do not have to send their children to school!*** (This is true for any child, not just brain-injured children). The 1944 Education Act gives every parent the right not to send their child to school, as long as parents do provide an ***appropriate*** education for the child.

The key word here is ***appropriate***. Obviously what is appropriate for seven year old Jimmy, who has no developmental problems, is not appropriate for a child with profound brain-injuries. As long as parents are able to demonstrate to the local education authority that they are working to improve their child's developmental problems, (such as carrying out one of our developmental programmes), they cannot lift a finger to force you to do anything. They may wish to send along a plethora of 'professionals' to give you advice; - that's fine! – You are under no obligation to take that advice. They may try to persuade you to set up 'links' with your local special school, or to just go in there for a few hours a week to 'receive advice;' – Be aware of your rights. You do not have to do this if you do not want to!

There is one thing of which we would advise you to be very careful! Once the name of a child who has 'special educational needs' is placed on the register of a school, it can be a legal nightmare to try to 'de-register' him. – It takes the personal involvement of the secretary of state for education to do so! So, if you do not wish to have anything to do with the school system, as many parents do not, be careful about any involvement, no matter how minor, with school. Be clear that you do not wish for your child to be registered with the school and ensure that you receive an acknowledgement *in writing* which gives the assurance that this will not take place. In addition, if you do choose for your child to receive a statement of special educational needs, it is a good idea to include on that statement your wishes for your child not to attend school. You are perfectly within your rights not to send your child to school; - the 1944 Education act guarantees you those rights.

Some professionals can be very clever at 'boxing in' parents by using a statement of special educational needs as a coercive tool. Always ask for copies of any reports, which professionals may write and make sure that you give your views on the content of those reports in writing, asking them also to include your views within their own reports. It is also a good idea periodically to access your child's medical records, which are available at your local hospital. This will ensure that no professional is expressing views and opinions without you being aware of it. Your child's doctors and other concerned professionals may be uncomfortable with this and you may not be popular, but this is not a popularity contest, you are fighting to give your child the best possible chance in life!

My second disagreement with 'statementing' arises when a situation occurs where there is an attempt to 'statement' a child who is well under school age. I am aware of professionals who have attempted to do this with children as young as two years of age, just as they did with my own son. Why would they wish to do this? How can anyone conceivably attempt to determine the educational needs of a child who is so young? School is still years away and there is no way of being sure of how well or badly the child's development will progress! This is merely another tactic, which some unscrupulous people will use to try to 'persuade' parents that the professionals' wishes for the child are best. At this tender age, parents are still relatively inexperienced concerning their children's problems and delivering a process such as 'statementing' may sway the parents towards the opinions of the professionals. Quite simply, you have the right to tell them to go away and to come back when your child reaches school age.

Feeding difficulties

Many children who have brain-injuries, especially those whose problems are more severe, have difficulties with feeding: This is due to various factors:-

- The breathing pattern of many children with brain-injury is shallow and fast. It sometimes never develops from that of a newborn. This type of breathing is not conducive to effective eating. Try running up and down the stairs a few times and then attempt to eat whilst you are out of breath. This is the trauma, which many of our children face at every meal-time.
- Children with brain-injury may also have difficulty in coordinating their breathing in conjunction with chewing and swallowing. This can be particularly difficult when you are not feeding yourself and someone else is dictating the pace of the feeding.
- The swallowing reflex of many children with brain-injury is not as efficient as it should be. In these cases, children can aspirate, (breathe in) their food, which can be a contributory factor in the development of pneumonia
- Many children who suffer from brain-injury do not possess the necessary muscular coordination to chew effectively and therefore have difficulty eating solid foods.

Constipation

I apologise to the reader for speaking in such graphic terms about such a subject, but this is one of the most common and important problems, which brain-injured children face. Constipation can exacerbate many other symptoms of brain-injury, such as epilepsy, poor sleep pattern, irritability, muscle tone and general responsiveness just to highlight a few.

Often it is difficult to spot this problem, especially if you are unaware that it *is* a problem of any significance. In many instances, no one alerts parents to this potential difficulty and consequently they are largely unaware until it is too late. The professional approach in many cases seems to be to wait until a problem arises and the child is admitted to hospital before any information is imparted to parents. Very often at this point, with parents suffering extreme anxiety, the doctor will say, as he did to one parent to whom I spoke, *"Oh it's not your fault, this happens all the time with these children."* May I politely point out that maybe, just maybe if parents were alerted to potential difficulties then they would not *'happen all the time!'*

One particular child with whom I was involved had given no direct indication that constipation was a problem. He was passing stools regularly, so much so, that it was noted that his bowel function was showing great improvement. What was actually occurring in the case of this child, which is mirrored by many, many other children is that some faeces had impacted in the intestine and was causing a 'roadblock' which was 'backing up' the entire system. The reason the young lad was passing stools regularly and consequently bamboozling his parents, was that the pressure of the faeces behind the roadblock was so great that faeces was being forced around the sides of it. This was the faeces he was passing and to their untrained eyes, his bowel function was normal. The truth was that he was in terrible difficulty.

What these parents were now told by the doctor highlights what I have just said about lack of information imparted to parents and it caused them to be very angry indeed. *"We asked the doctor if in some way we were at fault in allowing this to happen. Her response was; 'You were not to know. I'm surprised he's not been in hospital before!'"*

If the parents had known that constipation of this severity was a possibility, they would have taken steps to avoid it. In the ensuing years, they were able to ensure the problem did not recur. A little information can be a powerful tool!

Surely when a brain-injured child is born into a family, once the family has recovered from the shock, they need to be furnished with as much information as possible, not kept in the dark and above all not treated like imbeciles! Parents are at the battlefront in the fight against brain-injury and very often, because they are not provided with relevant information, they are forced to fight blindfolded and without the necessary weaponry!

Over a period of time, families need to be educated about the everyday problems, which can arise with children. It would save many children a great deal of pain and discomfort, because parents would then have the knowledge to spot an impending problem and bring it to the attention of the doctor. This would also save the health service a great deal of money and would save parents untold anguish.

Allow me through this chapter, to make parents aware of constipation. One of the primary reasons why our children can become constipated is immobility. A highly active, mobile child will usually have no difficulty in this direction, whereas brain-injured children often experience problems with mobility, in many cases having little mobility at all.

Another reason for constipation is that a bowel movement requires sophisticated muscular coordination. One of the primary problems faced by any brain–injured child is lack of muscular coordination.

A third factor is inadequate liquid intake. We all need a specific amount of liquid in order for our bodily functions to run smoothly. Many children lack the sophisticated muscular coordination to be able to cope with swallowing liquids effectively and therefore do not consume them in the quantities, which are necessary for optimum health. There are many other reasons why the symptoms surrounding brain-injury are instrumental in causing constipation, but they are much too detailed to explore here.

Epilepsy

Epilepsy is a very common problem suffered by brain-injured children. It is very distressing to watch and can be debilitating for the child. The variation of seizures is incredible, - just when I think I have probably witnessed just about every variety of epilepsy, up pops a child who produces something unique.

What is epilepsy?

Epilepsy is the tendency of specific brain-cells to misfire. There are three distinctions of epilepsy to bear in mind in understanding this problem. The first distinction is that an epileptic episode is either ***partial*** or ***generalised***. The second distinction is that an epileptic episode is either ***simple*** or ***complex***. The third distinction is that epileptic seizures are either ***grand mal*** or ***petit mal***. Confused? – Don't worry, read on and it will all become clear.

Normally, brain cells fire according to their being excited beyond a certain threshold of stimulation, or they can be restrained from firing because they are inhibited from doing so. Sometimes in epilepsy, these inhibition and excitation thresholds are not applied successfully, causing brain cells to misfire very easily. When this happens, two things can occur:

- The misfiring may be limited to a specific area of the brain, causing a very specific response from the individual such as a short ***absence*** or a twitching of one limb. These are known as ***partial*** seizures.
- The misfiring may form a chain reaction, which spreads to a larger area of the brain, causing a more ***generalised*** response from the brain. Adams & Victor (1981), successfully demonstrated this phenomenon by measuring seizure activity with electrodes placed inside patients' brains.

Partial seizures can be further subdivided by the second distinction of ***simple*** and ***complex***.

Simple, partial seizures bring about changes in the level of consciousness, but never involve a loss of consciousness, whereas ***complex partial seizures*** do involve a loss of consciousness.

Sometimes, if the focus of the epileptic activity is in one of the ***temporal lobes*** of the brain, the child may experience an ***aura*** prior to the attack. This 'aura' may be an experience of positive or negative emotions, it may be a hallucination of one or more sensory modality, or the aura may trigger memories or stereotypical movements.

We now come to the third distinction of seizure activity, ***grand mal*** and ***petit mal*** seizures. Sometimes during a more dramatic 'complex partial' seizure, the child's body may rhythmically shake. This is known as a ***grand mal*** or ***tonic - clonic*** seizure. Although this looks dramatic, it is nothing to be alarmed about and is usually over within a few minutes as a combination of structures in the brain, collectively known as the ***diencephalon*** act to suppress the seizure activity.

A ***petit mal*** seizure is less dramatic, usually very brief and is sometimes so shallow as to go unnoticed by parents, teachers and doctors. An example of such a seizure would be an ***absence***, where the child simply stares vacantly for a second or two. I know a child who used to experience dozens of such absences an hour and although they sound unobtrusive, the disruptive effect they can have upon life is easily underestimated.

Finally, and importantly, one aspect of epileptic activity it is important to discuss is a phenomenon known as ***status epilepticus***. Usually seizure activity will dissipate within a few minutes and the child will recover with no harm done. However, rarely the seizure activity either will not stop, or the child emerges from one seizure to quickly be consumed by another and another. This is a dangerous and potentially life threatening situation, which needs immediate medical intervention. In extreme cases such as this, seizures are capable of causing further brain-damage. It has been demonstrated that some patients with seizure disorders display injury to a part of the brain called the ***hippocampus***, the amount of damage being closely correlated with the number and severity of seizures, which the patient has experienced. The damage appears to be caused by the excessive release of a neurochemical called ***glutamate*** during the seizure. (Thompson et al, 1996). It is consequently vital that even if you only suspect your child to be experiencing seizure activity, that this is checked out by a doctor.

What causes epilepsy?

Epilepsy has two causes, one is *pathological,* and the second is *physiological* in nature.

- Pathology: -When a brain suffers injury, millions of brain cells may die. Around the area of injury, there may also be cells, which have not been killed, but which are nevertheless injured. In addition, as I have previously alluded, the thresholds of excitation and inhibition, which normally control the firing of these cells, may have been disrupted. Consequently, these injured cells may not fire according to their normal patterns, but may misfire. The child in this situation may have more or less constant epileptic activity occurring in his brain.

- Physiology: - The environment in which the brain operates is by necessity oxygen and nutrient rich. Although the brain only comprises approximately 2% of the body's weight, it consumes 25% of the body's oxygen intake. When a brain suffers injury, the availability of the oxygen supply can be compromised. For instance, in many cases of brain-injury, the development of the rate and depth of breathing of the child does not proceed from that of a newborn, which is fast and shallow; - this places difficulties on the optimum levels of oxygen availability, thereby compromising the physiological environment of the brain. Similarly, uninjured children who are ill and develop a temperature, may suffer an epileptic seizure as the temperature rise deprives the brain of oxygen, temporarily creating a poor physiological environment.

The young child with brain-injuries may also experience difficulties in taking in adequate nutrition, which could have the similar physiological effects. The brain's response to this impoverished physiological environment is to produce a seizure reaction. When brain cells struggle to operate normally without the oxygen and the nutrition they need, they begin to misfire.

What can be done to combat epilepsy?

There are many approaches to combating epilepsy, which may improve your child's situation. Often a programme of developmental stimulation can reduce seizure activity. The reason for this is that if developmental gains can be produced in the child, it means the brain is operating at a more mature, efficient level. This 'brain development' can suppress seizure activity.

It is also important that children, who are predisposed towards epilepsy, be given good nutrition. This will help to maintain the physiological environment of the brain in as optimal a state as possible and hopefully keep seizure activity to a minimum.

There is a wealth of medical technology now available, which can help to combat seizure activity. Anti-convulsant medication always extracts a price from the child in terms of drowsiness and other side effects, but it is often necessary, if only temporarily to keep epilepsy under control. As long as the levels of medication are kept to the minimum levels needed to control the seizure, there is no harm in pursuing this path.

Ok, having reviewed all my ideas for treatment, you may decide that you would like to pursue this avenue. You may decide to ask us to prescribe a programme of developmental stimulation for your child, or you may decide to pursue another route. Alternatively, you may decide to do nothing at all: - This is entirely your decision and as I say, no–one should criticise you for it.

Whether your choice is action or inaction, many of you will ultimately take decisions for the benefit of your child, which will bring you into conflict with one or more of the medical / health professionals who are supposed to be there to support you. So let us look at parent – professional relationships.

Parent – professional and family relations

A major problem faced by many parents, whether they pursue alternative treatments or not, is the response to the family of the healthcare professionals. Many parents report negative experiences in dealing with professionals, especially if they are following a course of action with which the professional disagrees. One family reported an incident to us, which quite frankly is typical!

Their trouble began with a physiotherapist who regularly visited their home to treat their four-year-old son, who suffered from profound brain-injury. Soon after they had begun a programme of developmental stimulation, - a situation by which this particular therapist obviously felt threatened, she began to make derogatory comments about the fact that their son would occasionally incur a minor rub or bruise from trying to crawl down his wooden ramp. This is a situation, which is unfortunate, but it occasionally happens. It is usually remedied by a slight alteration in the angle of the ramp to the floor. The bruises were not frequent or severe and one would expect a well, uninjured toddler to pick up such bruises in the course of stumbling about the house. At first, the parents studiously made no response to the remarks, but the increasing malevolence of the words was building to a crescendo. The 'tipping point' occurred when, at the end of one visit the therapist disparagingly said *"I wonder what social services would make of those marks on Gareth's leg?"* This was simply too much for the family to take and they confronted the therapist, asking precisely what point she was trying to make. As is always the case with bullies, the moment they are confronted, or asked to justify their words, they 'turn tail' and run! As she quickly disappeared through the door, she mumbled something about 'joking' and not taking her seriously. She apparently never uttered another remark about Gareth's minor scrapes.

Yet another confrontation between one of my clients and the ranks of the 'professionals' emerged when a new paediatrician asked a particular family to pay her a visit in her office, as she wished to get to know their daughter and become acquainted with her problems. (at least this was what the doctor stated on the telephone!) She gave the impression of being very supportive towards the family and appeared to be genuinely interested in what they were trying to achieve for their daughter. Suitably lulled into a false sense of security, they happily took their daughter, Julie along to see the doctor. However, when they arrived in her office, she told them plainly that she fundamentally disagreed with what they were doing in terms of therapy and what was more, were they to continue, she would *"not touch Julie with a barge pole!"* The family simply could not believe the complete turnaround in the woman's attitude and asked her to tell them precisely what it was about Julie's therapy that offended her so much. Of course, as has been the case with every single member of the medical profession I have encountered, she was unable to substantiate her view with any evidence. - She was unable to say what she found disagreeable because she did not actually **know** anything about what it was she so violently disagreed with!

Again, as is typical of many 'professionals' I have encountered, she was merely reacting to rumour and hearsay generated from within her profession and being the unimaginative sheep, which some of them undoubtedly are, their opinion must mirror that of the flock! Why allow truth, fact and evidence to spoil good prejudicial opinion?

The woman then continued to accuse the family of being ***"on a high"*** stating that her perception was that they seemed to be ***enjoying*** their daughter's disabilities! What kind of twisted, warped personality could imagine that parents would enjoy their child's difficulties? In this particular case, all the family were trying to do was to undo the harm, which had been inflicted upon Julie through medical negligence at birth; - medical negligence fostered by just this kind of conceited arrogance.

The most horrific incident, of which I became aware, involving a family and a 'professional' was played out in the absence of that family, but was reported to them by a friend. Just prior to their son beginning a programme of developmental activities, the family in question had moved away from their old neighbourhood to an area approximately fifteen miles from where their son had been born. Obviously, via the 'professional' grapevine, news of the family embarking upon a different type of therapy had travelled back to where they formerly lived and other parents of children with developmental disabilities, who attended the same clinics, which the family formerly attended with their son James, began asking questions about the family and their new programme. One of these parents, with whom the family had become friendly during their clinic visits, telephoned one evening in great concern to ask if the family were both coping with the ***divorce proceedings!*** Of course, no such event was occurring at all! Apparently, desperate that other parents did not follow in the footsteps of this family, one 'professional' had elected to inform the other families that the new programme on which the family had embarked was so stressful, it had broken up their marriage and they were in the process of going through a particularly acrimonious divorce! The friend was assured that this was not the case, that although the programme was indeed hard work, their relationship was more solid now than it had ever been.

To this day, I have difficulty believing that an adult human being, particularly one claiming to be a 'professional' in the caring services, would be so vicious and nasty. I feel these remarks were made out of fear; - a fear borne out of inadequacy at having no solutions to the difficulties our children face. A fear borne out of viewing a therapeutic intervention, which was outside their jurisdiction as a threat to their professional competence!

The professional attitude towards parents reported by parents themselves is one, which appears to be peppered with arrogance. Parents, instead of being viewed as partners and valuable information sources more often feel as though they are viewed as an integral part of the problem. Unfortunately, this means that whatever parents say will not be taken seriously, if it is listened to at all! It is probably this display of arrogance and contempt, which is instrumental in parents wanting to seek 'alternative' solutions to their children's problems in the first place. In other words doctors, it is often *your* attitude, which drives families away from you!

Yet another 'showdown' was reported to me when another doctor called at the home of one of my families to perform a routine, periodic developmental assessment on a child. According to the family. during his visit he did his level best to explain in his 'doctor knows best' voice, what a terrible mistake the family were making by following the developmental programme I had devised for the child. He stated that, in his opinion, this type of therapy was scientifically unfounded and simply would not work. The family asked him what it was precisely about their son's programme that he objected to. – Surprise, surprise, he could not provide an answer to this question as he possessed no real knowledge of what the family were doing.

He was then provided with several research papers from eminent psychologists, verifying the techniques within their son's programme, which demonstrated the validity of the approach. Faced with hard evidence, provided by reputable academics, this buffoon still tried to argue that the family were wasting their time! Notably, unlike the family carrying out the programme, he provided no evidence to substantiate *his* argument.

To the family's credit they pursued this point further by asking him what he thought of the prospects for their son if they decided not to do anything to help him at all, instead being prepared to leave him trapped in a body which he couldn't control? The parents stated that in their eyes they considered *that* approach to be cruel! Staggeringly, his response was that their son was happy enough now, so why not leave him? – This attitude says it all! I tell you this, although I harbour no criticism of parents who decide not to follow any treatment methodology with their child, there is one inescapable truth to be told: - ***Doing nothing is likely to achieve nothing!***

Surely to be a scientist, (which so many health professionals love to claim to be) means that one should approach a situation with an open mind and only be swayed by evidence; - instead this man was listening and subscribing to prejudice and hearsay.

When a family embark upon an alternative therapy programme, as long as the approach cannot cause harm to the child, their decision should be respected. Parents do not make this type of choice because they wish to elicit open confrontation, or with the aim of excluding treatment and advice from health professionals. Parents should not have to consider that anyone might be *upset* by their decision to turn every stone to help their child. This however is the way in which health care professionals often seem to interpret such decisions. The very people, who even if they disagree with parents, should continue to offer their support and advice, often react with nothing short of hysteria and open hostility.

In another instance, one family asked why their paediatrician had been so offensive about their chosen course of therapy for their son, Mark. The doctor replied that she disagreed with the intensity of the programme and had seen marriages broken apart by such regimes. The family pointed out that they were only working with Mark for three hours per day, which was all the time they could commit and that the activities in which they were involved were actually more akin to play sessions and were not intensive at all. They highlighted that if Mark attended school, which he could, he would then be exposed to a six - hour day educational programme. The doctor simply would not listen and the encounter ended with them asking to be referred to another doctor.

Some families become tired, bored and frustrated with the constant hostility which some 'professionals' throw in their direction. Apparently some of them act in this way because in trying their best to help their child, parents are not following the doctor's wishes. So it seems, in some cases it may be a matter of loss of power and control, not medical concern which are the motivating factors behind medical reaction. Most parents have enough to do and enough problems to deal with, without having to put up with narrow minded criticism from so called 'professionals' who seemingly possess few solutions themselves!

In my opinion, this doctor was fundamentally wrong; - a marriage would be more likely to encounter difficulties if the couple were forced to sit back and merely watch as their child simply did not develop at all, or even deteriorated.

It is clear by the contents of the letter written by this doctor after her visit, that what was related to her about the intensity of Mark's programme, was simply not comprehended!' This was no surprise to me at all, as I have realised from bitter experience, parents simply are not listened to!

In her letter she writes:-

"When I saw them, I expressed my extreme concern about this type of intensive physiotherapy."

Well, for a start, if she had actually read the information provided for her prior to her visit, she would have realised that the programme was absolutely nothing to do with physiotherapy, and as was explained to her it was not intensive at all!

It is clear to see in the academic literature and it is a view, which has been reinforced by every parent to whom I have ever spoken, that the major source of stress which families face is not their child's disability, but the relationship with the people who should be there to support parents: - the healthcare professionals!

As far as marital problems are concerned, there is no evidence that embarking upon a particular type of therapy with a child is likely to place a marriage under any extra duress. The most prevalent factor within the situation seems to be pre-existing stress between all members of the family, which may adversely affect family relationships. This is not caused by a type of therapy, but by the stresses caused by the child's problems. A child who has brain-injuries, who does not sleep, has problems taking in sufficient sustenance, is difficult to handle physically and has associated behavioural, toileting and medical difficulties, will not create a situation where parents are calm relaxed and happy!

The difficulties brought about by the child's brain-injuries, (note, *not* by any form of therapy), can affect the way in which the parents not only react to one another, but to other family members. Most authorities agree the presence of a child with brain-injury within a family most often has the effect of 'cementing' an already strong relationship, or breaking a 'weak' one. (Thomas & Swann 1987; Worthington, 1982).

One leading paediatrician to whom I spoke confirmed this view. He claimed after a few months of getting to know the family, he could usually make an accurate guess as to which relationships would not survive the ordeal. He asserted that in his experience, the presence of the child intensifies the pattern the relationship already has.

Indeed, it is unfortunate that the problems which are posed by the child's difficulties (not a therapeutic intervention), can cause considerable marital conflict and even, in some instances divorce! One of the foremost difficulties seems to be that parents simply do not have as much time to devote to each other, because the child's problems are so demanding. (Clarkson et al, 1987).

So, reviewing the actual *evidence,* it is the *child's condition*, which causes marital stress, not any particular *therapeutic intervention,* which the parents may choose to follow with their child.

The *evidence* refutes the claim of this particular doctor that programmes such as that prescribed by Snowdrop and other clinics cause untold stress within families. It is simply not the truth! According to the *evidence*, such stress is caused by the child's condition and the parents' relationship with healthcare professionals! Two studies in particular demonstrate that mothers of children with brain injury, (studies in which the children are not following any particular form of treatment!), suffer significantly more psychiatric symptoms than do 'ordinary' mothers. (Clarkson, 1987; Hirose, 1990).

Surely, a developmental programme of the type we are discussing, would help to alleviate such stress, because parents would at least feel that they were trying to achieve something positive. They would be more likely to feel that they were productively trying to help their child.

The problems, which parents encounter due to the child's condition, such as having time for each other are also transferred to interaction time with other children. Consequently, siblings may feel neglected and show jealousy towards their injured brother or sister. (Furneaux, 1988; Andersson, 1988)

These are the genuine difficulties, which have been uncovered by academic research, they are not problems invented by the wishful thinking of a doctor, who quite simply merely wishes to stop you seeking valid alternatives for your child.

So, contrary to medical opinion, Snowdrop type programmes are likely to decrease the social isolation and stress felt within families of children with brain-injuries.

An interesting study, commissioned in the nineties, cross–referenced the psychological well being of siblings with the amount of interaction between them and their brain–injured brother or sister. The results of this research supported the theory that children who are encouraged into greater interaction with their injured sibling are more self–confident than children who do not receive such encouragement. It was also noted that there was an overall improvement in family interaction patterns and a functional change for the better in the child suffering from the brain–injury. (Craft, 1990). Kind of an argument for involvement in some sort of programme isn't it?

Another major concern, which for most parents causes untold heartache, is the reaction of other people to their injured child. Parents seem to pass through definable phases in their reactions to strangers, initially wanting to talk, to tell the world all about their child. Indeed, I remember being a complete and utter bore concerning my own son. Someone would only have to mention his name to elicit a three-hour lecture on the ills of brain-injury!

Of course, this is not only a sign that parents feel isolated, it is also a sure sign that they need someone to talk to. It is yet another unfortunate failing of our healthcare system that often strangers have to fulfil the role of counsellor.

Later it seems, parents wish to be left alone and even resent inquiring strangers. Again, I can vouch for this as it mirrors my own experience and the experiences of many other parents to whom I have spoken. As the years passed and the problems faced by my own son became more and more visibly obvious, people would naturally look at him. I never liked this, but do understand that it is difficult not to look at someone who appears so different. The problem, with which I always grappled, was that some people would come back for a second look, or make it obvious that they were scrutinising my son. Some people simply have no class!

All you need is love

I know I am stating the obvious here and I am not saying that parents do not love their children. Neither am I trying to tell parents **how** to love their children. Parents are already experts at that and they do not need my help or advice. However, it is very easy to lose sight of your child as a person when you are immersed in the purpose of trying to solve your child's problems and you are planning and carrying out a schedule of developmental stimulation. As the old aphorism states: - *It is very easy, when up to one's ears in Alligators, to remember that one's original intention was to drain the swamp!*

Should you decide to embark on any type of therapeutic intervention, you should consider carefully just what you are trying to achieve. Brain-injury cannot be cured! It certainly cannot be cured by me and I do not know of any other clinic anywhere in the world that would make such a claim. If they do, you should run out of the door as fast as possible! What we can do between us, parents and therapist working as a team, is to provide the best possible neurological environment for your child, which will enable him to make the maximum degree of progress. In doing so, care has to be taken that a situation is not created, which is making too heavy a level of demand upon the family. If this situation is allowed to develop, the programme becomes absolute drudgery for both the family and the child. The quality of stimulation being provided is then very poor. This kind of situation is also one where the identity of the child can be lost. The child can become almost an object, to which things are done! I propose that this is counterproductive not only to the child's developmental opportunities, but that it places unnecessary stress upon already overburdened families. If you ever feel you are losing sight of the wonderful individual underneath all the developmental stimulation you are providing; - your child, then you are trying too hard and could create a counterproductive environment. In this case, it is time for a re-think!

The most important aspect of your child's life is **YOU**. The relationship you build with your child, the love you give him and the love you receive back, will build a platform on which those initial interactions, which are the precursors to so much developmental progress, can begin.

My philosophy on the amount of daily time, which a family should spend in trying to stimulate their child is simple and depends upon two factors.

1. It depends upon the child. Should we attack the problems of a hyper-sensitive, over-anxious child with a day-long schedule of activities? Of course not! All we would be achieving would be to drive the child further away from social contact and further into himself. The amount and type of developmental stimulation prescribed needs to be carefully judged in order not to overwhelm the child. It must also be aimed at trying to create an enjoyable experience for both child and parent.

2. Not all families can accommodate a schedule, which takes a great deal of time, for various reasons. Economically families may not be able to stay afloat if they were to make sacrifices to accommodate such a schedule. There are a hundred reasons why a family may not be suitable candidate for a lengthy schedule of activities. Does this mean we should completely deny that child any developmental stimulation, which may change his life for the better? Of course not! We should work with each family to decide what is achievable in practical terms and in terms of time and effort.

What we are trying to achieve is 'happy families.' Only if the family is happy, relaxed and fully understanding of what they are trying to achieve, will the injured child respond in a manner, which is likely to encourage the building of his relationships with other members of the family. Only then, through those relationships of trust, mutual affection and the enjoyment of interaction can we begin the journey down the developmental pathway.

Where do we go from here?

After many years as a parent of a profoundly brain-injured child, almost as many years in academic research and working in schools with children who have special educational needs, I decided the time had arrived to fulfil a long held ambition. Having travelled internationally to clinics, both as a parent and as a researcher, I have become aware of just how little of the vast amount of knowledge possessed by the discipline of psychology is being applied to the treatment of children with brain-injuries. Knowledge from great psychologists such as Vygotsky and Luria; from Bruner and Rogoff and from a hundred other names. The knowledge, some of which is highlighted in this book, is out there waiting to be used for the benefit of our children. I have elected to try to utilise this knowledge and apply it to the treatment of brain-injury.

I have personal experience and knowledge of the serious consequences for the family, which the presence of a child with brain-injury can bring, not least of all the problem of stress, which family members bear. As I have previously stated, a child with brain-injuries, with his myriad of associated difficulties, is likely to create tired, stressed parents, - Understandably so! This needs to be taken into account when dealing with the family and in this sense, we are not merely treating the brain-injury, nor merely the child; - We are treating the whole family and if we are to have a significant impact upon the problems presented by brain-injury we must surely do so!

With these thoughts in mind, and having worked alongside parents and brain-injured children for many years, I have established a child development consultancy called 'Snowdrop.' (Snowdrops have a special significance for my family and our little boy. In symbolic terms, they also traditionally signify hope in adversity. - A symbol which reflects the status of many families). Although we are still young and small, we are treating children from all over the world, with a great deal of success.

My personal crusade will be to ensure that Snowdrop remains 'research friendly' and eclectic. By this, I mean we should not develop a set philosophy or rationale, which will effect our methodology. – Down that road lies academic egoism and inflexibility, which I have witnessed at so many other institutions and which would lead to stagnation. We should always look to research and to evidence to provide our treatment methods and not be afraid to change.

It is my hope that this book has been informative. It is not meant to be an extended text, which delves deeply into all of the problems faced by our children and their families; - There are so many problems that it would be unwieldy and impractical to present to parents. It is meant to give information to parents on the major issues, which they and their children will face. On that basis it was intended to be relatively brief and to the point! I hope it succeeded. Should you wish to contact Snowdrop, feel free to do so using the information at the end of the book.

References

Adams, R. D. & Victor, M. (1981). *Principles of Neurology.* New York. McGraw –Hill

Anderrson, E. (1988). Siblings of mentally handicapped children and their social relations. *Mental Handicap Bulletin. (71),* 5

Anastasiow, N., (1979). Oral Language – Expression of Thought. *Urbana III. ERIC International reading association.*

Baddeley, A. D. and Lewis, V. J., (1981). Inner active processes in reading. The inner voice, the inner ear and the inner eye. In Eysenk, M. W. and Keane, M. T. (1997). *Cognitive Psychology. (3rd Edition).* Hove. Psychology Press.

Banks, M. S., Aslin, R. N., and Letsin, R. D. (1975). Sensitive period for the development of human binocular vision. *Science, 190,* 675-677.

Beck, A. T., and Guthrie, T. (1956). Psychological significance of visual auras: Study of three cases with brain damage and seizures. *Psychosomatic Medicin,* Vol XVIII, no 2,

Berk, L. E. (1997). *Child Development.* (4th Edition) London. Boston. Allyn & Bacon.

Bettleheim, B. (1967). *The empty fortress: Infantile autism and the birth of self.* New York. Free Press.
Bobath, K., (1980). *A neurophysiological basis for the treatment of cerebral palsy.* London. Spastics international medical publications.

Bornstein, M. H. (1988). Mothers, infants and the development of cognitive competence. In Rogoff, B. (1990). *Apprenticeship in Thinking: Cognitive Development in Social Context.* Oxford. Oxford University Press.

Bruce, D. J., (1964). The Analysis of Word Sounds. *British Journal of Educational Psychology. 34.* 158-170

Bruno, L. (2006). *Conscious Living: How a Brain Injury Changed the Way I Make Decisions.*
http://www.internationalrenaissancecoaching.com/s_140.asp

Bryant, P. and Bradley, L. (1996). *Children's Reading Problems.* Oxford. Blackwell.

Carlson, N. R. (2007). *Physiology of Behavior.* London. Allyn and Bacon.

Clarkson, S. E., Clarkson, J. E., Dittmer, I. D., Flett, R., Linsell, C., Mullin, P. E., and Mullin, B. (1987). Impact of a handicapped child on the mental health of parents. *Mental Handicap Bulletin,* **(65),**

Collins, A. L. Deqiong M., Patrice L., Whitehead, E. R., Martin, H. H., Wright, R. K., Abramson, J. P., Hussman, J. L., Haines, M. L., Cuccaro, J. R., Pericak-Vance, G., and Pericak-Vance, M. A. (2006). Investigation of autism and GABA receptor subunit genes in multiple ethnic groups. Neurogenetics *(7),* 3, July 2006.

Corkum, V., & Moore, C. (1995). Development of joint visual attention in infants. In C. Moore & P. Dunham (Eds.), Joint attention: Its origin and role in development. Hillsdale, NJ: Erlbaum.

Corson, D. (1988). *Oral Language Across the Curriculum.* Clevedon. Multilingual Matters.

Craft, M. J., Lakin, J. A., Oppliger, K. D., Clancy, G. M. and Vanderlinden, K. W. (1990). Siblings as change agents for promoting the functional status of children with cerebral palsy. *Developmental Medicine and Child Neurology.* **(32).**

Deloach, J. S. (1983). *Joint picture book reading as memory training for toddlers.* Paper presented at the Meeting for the society for research in child development. Detroit.

Dunn, J., Brown, J., & Beardsall, L. (1990). Family talk about feeling states and children's later understanding of others emotions. *Developmental Psychology. 26.*

Ehrich, J. F. Vygotskian Inner Speech and the Reading Process. *Australian Journal of Educational & Developmental Psychology Vol. 6*, 2006, pp 12-25
Elbert, T. (1998). Brain Plasticity. http://www.abc.net.au/rn/talks/8.30/helthrpt/stories/s10302.htm

Furneaux, B. (1988). *Special Parents.* Milton Keynes. Open University Press.

Haist, F., Adamo, M., Westerfield, W., Courchesne,E., and Townsend, J., (2005). The functional neuroanatomy of spatial attention in autistic spectrum disorder. *Developmental Neuropsychology, 27, 3,* 425-458.

Henderson, B. B. (1984a). Parents and Exploration: The effect of context on individual differences in exploratory behaviour. *Child Development. 55* 1237 – 1245.

Henderson, B. B. (1984b). Social support and exploration. *Child Development 55,* 1246-1251.

Heywood, C. A. and Kentridge, R. W. (2003). Achromatopsia, color vision and cortex. *Neurology clinics of North America.* **(21),** 483-500.

Hirose, T., Veda, R., (1990). Long term follow up of cerebral palsied children and coping behaviour of parents. *Journal of Advanced Nursing*, **(15)**

Horne, J. A. (1978). A review of the biological effects of total sleep deprivation in man. *Biological Psychology.* **7**, 55-102.

Johnson, M. H., Posner, M. I., and Rothbart, M. K., (1991). Components of visual orienting in infancy: Contingnency learning, anticipatory looking and disengaging. *Journal of Cognitive Neuroscience, 3,* 335-344.

Josefi, O. and Ryan, V. (2004) 'Non-directive play therapy for young children with autism: A case study'. *Clinical Child Psychology and Psychiatry,* vol. 9, no. 4, pp533-551

Kleitman, N., (1961). The Nature of Dreaming. In, Wolstenholme, G. E. W., and O' Connor, M. (1961). *The Nature of Sleep.* London. Churchill.

Kleitman, N., (1982). Basic Rest – Activity Cycle. – 22 years later. *Sleep, (4),* 311-317.

Lewis, M., & Brooks-Gunn, J. (1979). *Social cognition and the acquisition of self.* New York. Plenum. In Schaffer, H. R.. (1996). *Social Development.* Oxford. Blackwell.
Littleton, K. (1998). *Cultural worlds of early childhood.* London & Milton Keynes. Routledge.

Macey, P. M., Henderson, L. A., Macey, K. E., Alger, J. R., Frysinger, R. C. Woo, M. A., Harper, R. K. Yan-Go, F. L., and Harper, R. M. (2002). Brain Morphology Associated with Obstructive Sleep Apnea. *American Journal of Respiratory and Critical Care Medicine.* **166**: 1382-1387

Marshall, J.C. and Newcombe, F. (1973) Patterns of paralexia: A psycholinguistic approach. Journal of Psycholinguistic Research, 2, 175-199.

Mercer, N. (1995). *The Guided Construction of Knowledge: Talk amongst teachers and learners.* Clevedon. Multilingual Matters.

Mulleners, W. M., Chronicle, E. P., Palmer, J, E., Koehler, P. J., and Vredeveld, J. W. (2001)
Suppression of perception in migraine: Evidence for reduced inhibition in the visual cortex
Neurology, January 23, 2001; 56(2): 178 - 183.

Murray, L., & Trevarthen, C. (1985). Emotional regulation of interactions between two – month olds and their mothers. In Rogoff, B. (1990). *Apprenticeship in thinking.* Oxford. Oxford University Press.

Nind, M. and Hewett, D. (1994). *Access to Communication.* London. David Fulton.

Peretz, I., Gagnon, L. and Bouchard, B. (1998). Music and emotion: Perceptual determinants, immediacy and isolation after brain damage. *Cognition 68,* 111-141.

Pipp, S., Easterbrooks, M. A., & Brown, S. R. (1993). Attachment status and complexity of infants' self and other – knowledge when tested with mother and father. *Social Development.* 2, 1-14.

Poizner, H., Feldman, A. G., Levin, M. F., Berkinblit, M. B., Hening, W. A., Patel, A., and Adamovich, S. V. (2000). The timing of arm – trunk coordination is deficient and vision – dependent in Parkinson's patients during reaching movements. *Experimental Brain Research,* 133. 279 – 292.

Posner, M. I. and Cohen, Y. (1984) Components of visual orienting. In: *Attention and Performance: Control of Language Processes, vol 10*, Eds H. Bouma, D. Boulshuis. DG Erlbaum: Hillsdale, NJ.
Rizzo, M. and Robin, D. A. (1990). Simultanagnosia: A defect of sustained attention yields insights on visual information processing. *Neurology,* **40,** 447-455.

Rogoff, B., Malkin, C., and Gilbride, K. (1984). Interaction with babies as guidance in development. In Rogoff, B. (1990). *Apprenticeship in Thinking: Cognitive Development in Social Context.* Oxford. Oxford University Press.

Rogoff, B. (1990). *Apprenticeship in thinking.* Oxford. Oxford University Press.

Rogoff, B., Mosier, C., Mistry, J., and Goncu, A. (1993). Toddlers' guided participation with their caregivers in cultural activity. In, *Cultural worlds of early childhood.* London & Milton Keynes. Routledge.
Schaffer, H. R.. (1996). *Social Development.* Oxford. Blackwell

Scherzer, A.L., Mike, V., and Ilson, J. (1976). Physical therapy as a determinant of change in the cerebral palsied infant. *Pediatrics. 58.* (1). Pp. 47-51

Slater, R. (1995). *Our changing brain: Causes, effects and responses.* Milton Keynes. Open University Press.

Stepakoff, S., Beebe, B., and Jaffe, J. (2006). Mother-infant tactile communication at four months: http://www.isisweb.org/ICIS2000Program/web_pages/group1039.html

Sterr A. (1998). Changed perceptions in Braille readers. Nature 1998;391:134-135

Stern, D. (1985). *The interpersonal world of the infant.* . In Woodhead, M. Faulkner, D. & Littleton, K. (Eds). (1998). *Cultural worlds of early childhood.* London and New York. Routledge.

Tervaniemi M., Medvedev SV., Alho K., Pakhomov SV., Roudas MS., Van Zuijen TL., and Naatanen R., (2000) Lateralised automatic auditory processing of phonetic versus musical information: a PET study. Hum Brain Mapp. *Jun;10(2):74-9.*

Tharp, R. G., and Gallimore, R. (1997). *Rousing minds to life: Teaching, learning and schooling in social context.* Cambridge. Cambridge University Press.

Thomas, D., and Swann, W. (1987). *Family views.* Milton Keynes. Open University Press.

Thomson, S. M., Fortunato, C., McKinney, R. A., Muller, M., and Gahwiler, B. H. (1996). Mechanisms underlying the neuropathological consequences of epileptic activity in the rat hippocampus in vitro. *Journal of Comparative Neurology,372,* 515-528.

Tomasello, M. (1995). Joint attention and social cognition. In C. Moore & P. Dunham (Eds.), Joint attention: Its origin and role in development. Hillsdale, NJ: Erlbaum.

Trevarthen, C. (1995). The child's need to learn a culture. *Children and Society,* 9(1), 1995.

Vygotsky, L. S. (1978). *Mind in society: The development of higher psychological processes.* Cambridge, MA: Harvard University Press. Published originally in Russian in 1930.

Wood, D., Bruner, J. S., and Ross, G. (1976). The role of tutoring in problem solving. *Journal of Child Psychology and Psychiatry.* 17(2): 89-100.

Wood, D. (1986). Aspects of teaching and learning. In Woodhead, M.; Faulkner, D., and

Worthington, A. (1982). Coming to terms with mental handicap. London. Helena Press.

Wright, T. and Nicholson, J. (1973). Physiotherapy for the spastic child: - An evaluation. *Developmental medicine and child neurology.* 15 146-163.
Yang, T., and Maunsell, J. H. R.. (2004) The effect of perceptual learning on neuronal responses in monkey visual area V4. *Journal of Neuroscience, 24,* 1617 – 1626.

Zelazo, P. R., Zelazo, N. A., and Kolb, S. (1972). "Walking" in the newborn. *Science. 176. p314-315.*

Zentner, I., and Kagan, J. (1998). Infants perception of consonance and dissonance in music. *Infant Behavior and development. 21,* 483-492.

Contact Address.

If you are interested in using Snowdrop's services, please write in the first instance to: -

The Snowdrop Child Development Foundation

PO Box 85

Cullompton

Devon

EX1 1WP

Visit our website at http://www.snowdrop.cc

Email us at: info@snowdrop.cc

Please include your contact details and Snowdrop will then contact you concerning an appointment.